The Evolutionary Dynamics
of Complex Systems

Monographs on the History and Philosophy of Biology

RICHARD BURIAN, RICHARD BURKHARDT, JR.,
RICHARD LEWONTIN, JOHN MAYNARD SMITH

EDITORS

The Cuvier-Geoffrey Debate: French Biology in the Decades Before Darwin
TOBY A. APPEL

Controlling Life: Jacques Loeb and the Engineering Ideal in Biology
PHILIP J. PAULY

Beyond the Gene: Cytoplasmic Inheritance and the Struggle for Authority in Genetics
JAN SAPP

The Heritage of Experimental Embryology: Hans Spemann and the Organizer
VIKTOR HAMBURGER

The Evolutionary Dynamics of Complex Systems: A Study in Biosocial Complexity
C. DYKE

The Evolutionary Dynamics of Complex Systems

A Study in Biosocial Complexity

C. DYKE

New York   Oxford
OXFORD UNIVERSITY PRESS
1988

Oxford University Press

Oxford   New York   Toronto
Delhi   Bombay   Calcutta   Madras   Karachi
Petaling Jaya   Singapore   Hong Kong   Tokyo
Nairobi   Dar es Salaam   Cape Town
Melbourne   Auckland

and associated companies in
Beirut   Berlin   Ibadan   Nicosia

Published by Oxford University Press, Inc.,
200 Madison Avenue, New York, New York 10016

Oxford is a registered trademark of Oxford University Press

Library of Congress Cataloging-in-Publication Data
Dyke, Charles.
The evolutionary dynamics of complex systems.
(Monographs on the history and philosophy of biology)
Bibliography: p.
Includes index.
1. Social evolution.   2. Sociobiology.
I. Title.   II. Series.
HM106.D95   1988   304.5   87-20394
ISBN 0-19-505176-9

2 4 6 8 9 7 5 3 1

Printed in the United States of America
on acid-free paper

... la scienza invece, lavorando sempre, e lavorando sul tutto; modificandosi, ma insensibilmente; avendo sempre per maestri quelli che avevan cominciato dell'essere suoi discepoli, era, direi quasi, una revisione continua, e in parte una compilazione continua delle Dodici Tavole, affidata o abbandonata a un decemvirato perpetuo.

Alessandro Manzoni, *Storia della Colonna Infame*

Preface

I began this project in 1981. It had two roots. First, I was interested in genuinely dynamic models for social, and in particular economic, systems. The social sciences themselves seemed sadly lacking in such models, especially ones with a fair degree of robustness. So I thought that models might be found in evolutionary biology, and, with suitable modifications, used for the investigation of social systems. Second, Wilsonian sociobiology was attracting a good deal of interest and obviously had to be evaluated as a source of new models and research strategies for the social sciences.

So I began to study the literature of evolutionary biology with more directed purpose than I had before. I was immensely aided in this study by being invited to participate in the NEH Summer Conference on the Philosophy of Biology run by Marjorie Grene and Richard Burian at Cornell University in the summer of 1982. Not only did I learn a great deal at the conference, but I also met some people who have since become friends, collaborators, and teachers. They have proved essential as my guides to the biological literature.

I found that the field of evolutionary biology was in a period of constructive change. Challenges to the simpler versions of the neo-Darwinian synthesis were coming from several directions: molecular biology and related areas, critics of sociobiology, studies in the origin of life, and studies in paleontology and macro-evolutionary theory. In particular, I found a large and obvious gap between the pronouncements of the theoretical defenders of the synthesis, and the research strategies of workers obtaining results at the bench and in the field, whose sense of the complexity of the systems they were studying was much more sophisticated than the canonical version of the synthesis would have led us to expect. The very complexity of the first-order work cast doubt on the more far-reaching claims of, say, the sociobiologists. Hence, I felt, any contribution to be made by evolutionary biology to the study of social systems would have to be based on a full appreciation of the complexity of the systems studied. Consequently, as will be seen in what follows, I cast my lot with the first-order researchers rather than with the theoretical defenders of the synthesis. I also make use of the theoretical work of those interested in the thermodynamics of open systems far from equilibrium, where the most important empirical work is yet to be done.

Within evolutionary biology itself, the debates are often characterized in terms of whether the synthesis is to be expanded and complexified, or replaced by a new synthesis. In the light of advances over the last few years, I no longer think that this (rather bureaucratic) issue is of prime importance. My assumption is that everyone involved genuinely wants to understand evolutionary processes as a first priority, and the chips will fall where the solid research casts them. I do know, however, that the narrow versions of the synthesis, focused *exclusively* on natural selection as the sole mechanism of evolutionary change, are not adequate to be extended into the social realm. This will be argued in what follows.

Part of the problem of extending the methods and results of evolutionary biology into the study of social systems is that there are underlying problems concerning issues in the "philosophy of science" which tend to clog the transition. These problems concern traditional issues of explanation, reduction, and research heuristics. I have tried to assemble an array of decisions about these problems which allows research to go forward smoothly. In every case these decisions are underwritten by recent contributions from philosophy and from biology itself.

During the last few years quite a lot of good work has been done in the philosophy, sociology, and history of science by women, many of whom explicitly identify themselves as feminists. I have tried, with the help of an extremely able graduate student, Lynn Nelson, to keep up with this work, and have been particularly careful to write with it in mind.

Early versions of this book were much more polemical. As my priorities became clearer I found that what mattered most was what had led to the project in the first place: the search for productive ways of integrating research in the evolution of complex systems, whether they be biological or social systems.

As is usual with my writings, this book is built out of the hard work of others whose views, insights, and results I glue together. But, on the other hand, it is not the *Grand Synthesis:* the new "paradigm" to replace the old. First, as I argue throughout, a new synthesis will not emerge from the echoing vastnesses inside the head of a philosopher. It will emerge from the labs, field studies, and other investigative practices of serious investigators. The synthesis will be precisely the synthesis required to integrate the new knowledge they provide us.

The book seeks unities among things often thought to be different from one another; but it also insists on the differences among things often squashed into unities. Thus, any synthesis actually achieved will be far different from those envisioned, say, by the old positivist "Unity of Science" program. As a corollary, the sorts of links forged between biology and society will be different from those proposed by orthodox sociobiology. Will the unity of differences I try to put together work out in the end? Part of the thesis of this book is that there *is* no "in the end," in the sense of "final analysis". But we can still try to write books that advance matters for a time.

If this book does advance any matters, the credit will be due largely to David Depew and John Jungck, the silent but extremely influential partners in this enterprise. Only a half step behind in importance are Dick Burian, Evelyn Keller, Jeff Wicken, John Collier, Kyriakos Kontopoulos, and Joe Margolis. *Special* thanks go to Marjorie Grene. The actual production of this manuscript was made possible by contributions from Grace Stuart, Susan Meigs, Sol Lefkowitz, Juan Valdez, and our friends at IBM, Lifetree, and Buttonware.

Part of chapter 3 appeared in *Rivista de Biologia—Biology Forum* 79, 1986. An earlier version of material in chapter 4 appeared in *Evolution at a Crossroads: The New Biology and the New Philosophy of Science,* edited by David J. Depew and Bruce H. Weber, Bradford Books, MIT Press, 1985. Thanks are due to the respective editors for permission to represent that material here.

The book is dedicated to Duncan McDonald—naturalist in a world of biologists—who took up my education as an act of friendship, and taught me to appreciate the complexity, intricacy, and beauty of nature.

Doylestown, Penn.   C. D.
May 1987

Contents

The Evolutionary Dynamics
of Complex Systems

CHAPTER 1

Introduction

In extending the techniques and findings of contemporary evolutionary theory to social realms, sociobiologists have begun the strongest attempt to contribute to a coherent, overall view of human life as simultaneously biological and social. They have caught the attention of a wide reading public. Philosophers, social and natural scientists of all sorts, not to mention journalists and other observers of the contemporary intellectual scene, have all been stimulated to make their own comments and contributions to the ongoing debate—started, primarily, by E. O. Wilson's landmark work (Wilson 1975).

A perhaps unintended, but at any rate inevitable, effect of the sociobiology program is that all interested potential participants have been challenged to broaden their sphere of competence. The serious scientific credentials of the sociobiology program, not to mention its potential major contribution, make it mandatory for those of us who are not biologists to become as fluent as we can in the vocabulary and techniques of sociobiology. Conversely, as the biologists extend their scope to social phenomena, they must broaden and deepen their understanding of the phenomena that newly become objects of their research. Philosophers, for their part, have a whole lot to learn, as do the journalists.

These new obligations to achieve competence, in their turn, entail major pedagogical and heuristic difficulties for participants in the debates who seriously want to be understood. Nearly every page of a book like this one poses problems of exposition and requires that pedagogical decisions be made. These decisions would be impossible without aiming at a primary readership, leaving the expository strategies with respect to other readers to be dealt with in a secondary way. In addition, the assumption has to be made that *all* readers share the sense of obligation imposed by the project and are consequently willing to increase their

3

knowledge. Of all the faults in the recent discussion and debate surrounding socio-biology, the worst has been differential ignorance.

Expository Decisions

The main task of this book is to provide an explanatory framework and investi-gative heuristic broad and strong enough to support the integration of biology and social science. The centrality of this task accounts for many of the choices I have made. Hence, for example, I will not engage in as much direct polemic with the standard Wilsonian program and its direct successors as might be expected. When I do make direct criticisms, it will be as necessary ground-clearing or context-framing. Thus, while there will be occasion to consider aspects of the Wilsonian program in the context of discussions of reduction and determinism, many implicit contrasts between the view I present and orthodox sociobiology will be left aside in favor of the development of my positive points.

Similarly, much that I do will be related in some way or other to problems and issues within the traditional philosophical agenda. But a major decision has been to move these agenda entirely into the background. Philosophers will very often be able to find their bearings in their own terms, for there is surely a "philosoph-ical" dimension to what I will be doing. However, no philosophical issue will be discussed "for its own sake," but only as a contribution to the main project. Hav-ing made this decision, I try to compensate by providing some notes specifically designed to orient philosophers. This decision was made easier by my feeling that much of the traditional philosophical agenda is a waste of time—idle talk inher-ited from groping intellectual ancestors and by now obsolete.

Another expositional strategy is made mandatory by the serious consequences likely to flow from a well-founded and well-articulated integration of biology and the social sciences. In embarking on such a project, biologists commit themselves to a greater degree of self-consciousness about their work than they may be used to. So part of my project necessarily involves contributing to this self-conscious-ness. In moving into areas once reserved for the social sciences and philosophy, biologists are immersing themselves in an intellectual community far larger than the one in which they normally work. Over the last few years a number of socio-biologists have discovered this very fact, sometimes to their chagrin. But active participation in a serious intellectual context does impose its obligations—obli-gations such as those each of the disciplines insists upon in its own more limited context. And these obligations cannot be met unless everyone is fully, self-con-sciously, aware of the "rules and regulations" governing serious contribution, both in the home discipline and in the more extended one.

Part of the job of increasing self-conscious awareness of the nature and struc-ture of investigative disciplines consists in context building; although I do some of that, especially in the early stages, I keep such context-building "prolegomena" to a bare minium, in line with my decision to concentrate on a positive task. Nonetheless I realize that for some, especially biologists, even the bare minimum will be annoying. So, first, I ask that a clear distinction be maintained between discomfort and disagreement. (Another of my decisions is not to worry very much about discomfort, but to take disagreement very seriously.) And, second, I have a good deal of sympathy for the biologist who is impatient to get on with the substantive task. Such a person might well set the rest of this chapter and the next aside.

On the other hand, philosophers, social scientists, and others may find that this chapter and the next are key sources of disagreement, and that they are more tractable than some of the later chapters dealing with issues within evolutionary biology. In addition, the chapters in which I establish the initial features of my general framework allow me to pay my debts to those whose contributions I build upon. The task is far more integrative and synthetic than it is creative.

A convenient summary of my contextual considerations comes from Warren Weaver's oft-quoted statement of agenda (Weaver 1948, Peacocke 1983). Weaver said that the science of the Enlightenment taught us how to deal with organized simplicity. Nineteenth century science (Boltzmann, etc.) taught us to deal with disorganized complexity. The challenge for twentieth century science is to learn how to deal with organized complexity (without, I would add, pretending that it is simply conjunctive simplicity).[1]

I accept this as a manifesto, and in this introduction I try to provide an initial sense of some of the complexities to be confronted as we go along. Not only are the phenomena to be studied complex, but scientific practice itself is a phenomenon of organized complexity. The complexity of investigation must be studied along with the complexities investigated. The old positivist philosophy of science was a canon of simplicity, providing no room for a clear understanding of complexity. Insofar as working scientists (and social scientists) continue to understand their own activity in a positivist way (and many do), they will not find the space to meet Weaver's challenge.

The Political Economy of Evolution

We know that a key part of Darwin's worldview had its immediate source in Malthus. Darwin's debt to Malthus resulted in the now familiar conceptual connections between the theory of natural selection and orthodox economic theorizing. Since in the course of my argument I too will be providing links between biology and economy, we have to take a careful look at the links as they are forged within Darwinism. A synoptic sketch will be sufficient, for recent scholarship has established the major outlines.

A central feature of the Darwinian picture is that selection pressures are exogenous forces operating on the variation within a population at a given time. Secondarily, this variation itself may change due to mutations. These mutations are assigned to chance. Hence the only deterministic forces are those of selection. The forces of selection, in turn, are spelled out in terms of an economics of nature where scarcities of resources produce differential survival rates among variant individuals, thus modifying statistically the variation within populations. Natural selection, in other words, is a consequence of radical mismatches of supply and demand.[2] These mismatches are eradicated by fitting the demand to the supply, "restoring" equilibrium, the balance of nature. Thus, we might say, nature in its less than abundant bounty acts like an invisible hand to maintain a natural balance between available resources and the demand for them. Those organisms that survive are those best able to win the competition for the scarce resources. In the most gradualist versions of the theory the mismatches are small and continual, and the resulting restoration of equilibrium correspondingly continual, requiring but a tiny readjustment.

The theory, in short, is one of decentralized control of a complex natural system. The teleology so proudly eradicated by the Darwinian program is the teleology of a central planner with optimum design requirements in mind. As we will see later, other sorts of teleology are far less definitively eradicated. Right now we have to remember the reasons for eradicating the central planner. They were the desire to secularize nature to make it a fit object of scientific investigation, and the desire to combat mercantilism and the central—monarchical—authority it implies.[3] In fear of, and in opposition to, the claims of the would-be centralizers, both orthodox Darwinians and orthodox economists were led to affirm the diametrically opposite views.

The attempts of standard sociobiology to find "human universals" can be understood given the agenda of the early liberal political theorists. The conception operative in the specification of these universals is that in the absence of social constraints human behavior is handed over to the *default drive* of instinct. Thus knowing the *nature* of the default drive will allow us to predict the consequences of its operating unchecked. In other words, as has been pointed out quite frequently by now, sociobiology is at root a variation on a theme by Hobbes. To say this would not be to say much—everything has its historical antecedents—if it were not for the fact that the Hobbesian view of human life has been subjected to a great deal of critical scrutiny, which has unearthed the ways in which Hobbes' picture of humans fails to stand up to the standards of truth prevailing in science. Most of the critical judgments assessed on classical Hobbesianism are applicable with little or no modification to orthodox sociobiology.

The particular criticism that needs to be recalled here is a criticism not of the picture of human nature Hobbes and the sociobiologists provide, but of the presuppositions underlying the view that such a picture can be constructed. The logic of constructing a picture of human nature had perfectly understandable roots in Hobbes' time, namely Galilean and Newtonian roots. The operative analogy was with the eminently successful methods of abstraction which had established a viable physics. Just as the physical world, when all irrelevant characteristics (secondary qualities, etc.) were abstracted, was posited to be matter in motion, human beings, when all irrelevant accretions of culture were abstracted, were seen to be self-interested maximizers.

The problem with this view, by now exhibited many times, is that many analytical stripping-away processes are possible, and each yields its own conception of the default drive. The choice between them is not an easy one. It is not even clear that any of them is more than an exercise in fantasy. But historical reflection has shown how the Hobbesian program was a response to certain breakdowns and potential dysfunctions that were becoming evident as the modern era took form. People with a relatively stable morality, age-old ties to their place of origin, and stable (if modest) life expectations, were being uprooted, displaced, and their traditional social ties severed. For keen observers of this social dislocation two questions naturally arose. First, how would these uprooted people behave; and second, how would their resocialization have to be managed to make them upright members of civil society again? To theorists such as Hobbes and Mandeville it looked as if the way to get at this question was to ask what human beings would be, or would have been, in a state of nature, and then to rebuild on that basis.

The main mistake in the Hobbesian strategy from my point of view is to treat society and culture as overlays, adventitious accretions, instead of systematically integrated codeterminants of life. The bestiality described by Hobbes is a make-believe bestiality concocted of early Enlightenment fantasies not only about human beings, but about the beasts themselves. The state of nature so central to these theorists is more accurately the state of being driven out of society, with all the resentment and disappointment such a banishment entails. Now, at the time, given the social upheaval of the transition to modern society, this picture could well have been a useful one. Maybe the displaced and disaffected prowling the byways of England were best thought of as participants in the nasty, brutish, and short life Hobbes fabulates for them. But, if so, his picture is of a particular historical phenomenon with its particular determinants, not of some metaphysically primal state of Ur-humans.

The subsequent history of the Hobbesian strategy is charming, if not particularly convincing. From Locke to Kant to Rawls the transition was made to nicer, more "rational" Ur-humans and a correspondingly more pleasant state of nature for them to live in. Along the way, the fiction that the "state of nature" is a description of an actual or possible human existence was, as we know, gradually abandoned, until in the later theorists' work it functions purely as a normative ideal. In that guise it is used to establish conditions of "objectivity" and "rational" baselines for the grounding of the standard liberal exhortations. In its descriptive form it has been left to the speculations of the ethologists and sociobiologists, where its interpolation of liberal mythology bedevils attempts to integrate biological and social determinants of human life.

The Politics of Reduction

Science is an organized cluster of activities, recognized to be, like all human practices, embedded in a social matrix. Thus science is potentially subject to constraints imposed by the larger social whole of which it is a part. It is clear that this fact has to be resisted by science as a matter of policy. Indeed, Western science at least since Galileo has strained to be free of any such social constraints. It has pursued a combination of strategies designed (sometimes consciously, sometimes not) to win and defend its autonomy both as an activity and as an institution. It is in this light that reductionism can best be understood, for reductionism is not merely a normative canon of science. It is also a strategy for sealing science off from possible external constraints. It does so in the first instance by insisting dogmatically that the phenomena with which science deals are themselves immune to social influence—eternal objects long antedating human presence. The social inputs upon the objects studied by science are, in other words, set to zero.

For this assumption to be plausible, science must claim to be a totalizing activity, i.e., to be the sole legitimate source of the ultimate truth about the universe. Abstraction and reduction are performed to make such totalization possible (and plausible). In particular, physics has been extremely successful in establishing its program as the authoritative source of truth about the "fundamental nature of the universe." This success provides the impetus for attempts to reduce all other sciences to an atomistic base, for physics could then pass its authority "up the line" to other branches of investigative activity.

Reductionism assumes that the progress of science is detachable from the process of science. That is, it assumes that claims to scientific truth can be certified within the internal canons of science itself, but that the resulting certification commands standing in intellectual life as a whole. This is the megalomanic claim that otherwise cautious, modest, responsible scientists are forced to when they are dragged out of their labs to participate in the dialectic of legitimation taking place in the intellectual community at large. Science's conception of itself often is that it both has won the sole authority of TRUTH and has risen above the social context of debate. Neither is true (a simple empirical fact); and when this becomes obvious,[4] the lofty strategies adopted by scientists become dangerously dysfunctional.

In this light, science's defense of its activity in the public forum can be looked upon as an incidental activity only by accepting science's claim to autonomy *a priori*. Otherwise, it must be considered an integral part of scientific activity seen as one among many activities practiced in and for a society which has the right of intelligent review. A *sociobiology* which tries to seal itself off from the broader critical discourse would seem especially irresponsible.

It is clear that anyone who wishes to reduce human society and culture to biological determination has to argue that all structures constraining possible action—including the system of meanings which organize collective action and individual aspiration—can be articulated entirely in biological terms. Presumably this means that the presence of these structures can be explained on the basis of the biological substrate, and that once in place the structures do not take up a necessary role in further explanations of new structures or of actions performed under the constraints of these structures.

Obviously the biological reduction is unstable, since the extreme version of this strategy is physical reductionism, the ultimate stopping place for the reductive path. But the physicalist reductionist has an interesting problem. He wants to hold that all phenomena are reducible to the behavior of the physical substrate. Does he want to hold that his own theorizing is a phenomenon on a par with all others or that his own theorizing is special, and not subject to reduction? Of course if he chooses the second option he is no longer a reductionist, so we can assume that he chooses the first, and his own theorizing is, like every other phenomenon, reducible to the behavior of the physical substrate (Feigl 1958/1967; Margolis 1983).

But all physical phenomena we know (and we will see how difficult it is to stably demarcate "the physical") are undergoing change over time. The reductionist's account of the substrate to which all phenomena are to be reduced is itself, on his view, a physical phenomenon, hence subject to change. Hence his account is not eternal or necessarily immutable. Whether it remains the same over a period of time depends on its stability conditions within the successive environments it encounters. The *a priori* claim that a given account of the reductionist base is the *final* or *definitive* account is therefore untenable on reductionist grounds. An explanation for its eternal stability must be provided.

No such explanation has ever been provided, nor is it clear how one could be. Furthermore, any such explanation, if found, would prove the eternal stability *not* of a basic physical substrate, but of an extremely complex phenomenon, i.e., the belief that such and such *is* the ultimate physical substrate. Why should *this* complex phenomenon and not others be impervious to the vicissitudes of time? If it is, then it is hard to argue that it is a phenomenon on a par with other standard physical phenomena.

Close relatives of this argument are often found which optimistically conclude that physicalist reduction is incoherent, hence impossible. This conclusion is unwarranted, however. All that follows from the argument is that on physicalist grounds reductionism is an historical phenomenon subject to the same sorts of investigation as all other phenomena. Unless the reductionist can predict with certainty the future course of reductionist theorizing *and* the future results of scientific activity, indeterminacies and instabilities will be introduced into any nondialectical investigation of the nature of the substrate—that is, any investigation which does not simultaneously investigate the investigative process itself. These indeterminacies and instabilities can never be eradicated without denying the historicity of the conceptualization of the physical; *a fortiori* they can never be eradicated without denying the reduction of thought to the physical substrate. If reductive physicalism is stated nondialectically it denies itself.

Knower and Known

New strategies becoming well established in current philosophy of science reject the positivist claim that theories are to be evaluated solely on the basis of atemporal criteria. In the end this means that the investigative process must itself be treated as an investigatable subsystem with an internal dynamic and with relationships to other subsystems. The investigative process is both constrained by and constrains the objects of its investigations. From this point of view, methodology, epistemology, and ontology are all permanently susceptible to being put into question. Methodology I take to be the normative canons of research; epistemology the critical canons governing our choice of methodology; and ontology the set of objects of investigation presupposed or established by the methodology.

Perceived from this historical point of view, methodology, epistemology, and ontology are three interpenetrating organizers of investigative activity. Our methodological, epistemological, and ontological commitments constrain our investigations and prevent them from being anarchic guesswork. This is so even when we concede that our current commitments are themselves subject to criticism and revision. Our ability as a species to understand our world better and better depends on preventing our investigative activity from becoming random. Our methodology, epistemology, and ontology, provisional though they may be, have an essential role in bringing order to our attempts to learn and know.

To bring out this role I will emphasize a terminology which may appear clumsy at the outset, but which will serve us well as we proceed. I refer to methodology, epistemology, and ontology as structured structuring structures (Bourdieu 1977). They are *structured* since they are the ordered consequences of prior rational investigative activity and are "inherited" from this activity. They are *structuring* since they guide research and form the basis of further investigative strategies, constraining the range of what will count as further rational investigation, and both making possible and limiting what can be learned at any next stage. They are *structures* in an architectural sense which is barely metaphorical. Like stud walls and rafters, they are determinate organizers of possibility.

Thought of as structured structuring structures, methodology, epistemology, and ontology become part of a thoroughly secularized dialectic in which none of the constituents of the investigative process is immune from critique in the context of a constant structured restructuring. In addition, every investigatable subsystem—physical, biological, cultural, social, etc.—is a potential contributor to the restructuring of future investigative access. For example, some structural features of the social system within which orthodox evolutionary theory arose are partially responsible for the acceptance and stability (hence both success and limitations) of the Darwinian research program. Conversely, it is impossible to deny the effect that modern science has had on our conception of what there is in the world, or our conception of knowledge. Consequently we have to be careful of the way we locate Darwinism within broader social traditions. On the one hand, the pride felt by Darwinians in their contribution to human understanding is well justified. On the other hand, we have to be sensitive to features of Darwinism which are consequences of exogenous constraints imposed by the ambient social circumstances within which the theory was born and bred.

We also have to be careful to have the patience to work with the constantly dynamic structured restructuring of structures. Structures are almost always building up or breaking down, or in some other way being modified. The attempt to isolate and fix temporal thin sections effectively precludes perceiving this dynamic. Neither do structures, in general, have any essential identity-conferring features which necessarily persist through all their modifications. This means that if we cannot trace their historical path we may not be able to work with them at all. One of the advantages biologists have is that they are used to dealing with exactly analogous evolutionary transmogrifications within the earth's biota.

Science has to be seen as the activity of a complex system of "knowers" and "knowns," investigators and the results of investigations. Remarks about results are necessarily also remarks about the investigations producing those results. The history of modern science itself tells us that some results modify the way the process of investigation is conceptualized. At the same time, some reconceptualizations of investigative procedures modify the identification and articulation of results. We are always required to freeze ourselves in some historical slice of the investigative process, but we are also obliged to unfreeze and consider the process itself. Ultimate certainty falls in the light of our understanding of science as a human practice, i.e., as part of the subject matter of science itself.[5]

Once we attune ourselves to the historical state of scientific activity we are ready to focus on explicit programs of investigation. At any moment in ongoing research there will be a set of available models to apply to the phenomena being studied. Some of these models will have enough legitimacy to be live options. None of them will have enough legitimacy to be the only option. We know that we must pursue strategies that (a) are consistent with and legitimated by our earlier successful practice; (b) take full advantage of the resources and the models we have at our disposal; (c) do not foreclose any legitimate options that we might want reopened at a later stage; and (d) do not leave us with a tangled mess of hypotheses incapable of being integrated or even compared.

Thus we have a problem of research design on our hands much more complicated than the "hypothesis-testing" paradigm familiar in the positivistic socialization of the young researcher. In fact, research design ceases to be an *a priori* activity aimed at a pre-existing array of phenomena. Instead it becomes a coordinate and interpenetrating part of the investigatory problematic. In other words, the activity of investigation has to be considered as a whole, and any decisive delineation of investigator and object of investigation has to be based on boundaries which have been argued for, not arbitrarily set out beforehand. In general, the establishment of investigator/object boundaries will be partially a function of the models chosen as guides to research. The initial attempts to establish the boundary conditions will also be constrained, of course, by the success of previous research strategies.

As I have said, the progress of research depends on the process of research. This means that we have to look at first-order research itself to have a starting point for critically evaluating the methodology; and that critical evaluation must *at least* advance first-order research if it is to justify itself.[6]

So, to revert to the Weaver agenda, a framework has to be built for dealing with organized complexity. The explanatory framework the Humean and positivist traditions provide was designed for simplicity and linearity and has to be supplanted. But it cannot be supplanted by an explanatory framework incapable of accommodating the firm results obtained by the sciences of organized simplicity and disorganized complexity. Those results still stand as foundations upon which to build. So a reintegrated approach to explanation has to be formulated.

CHAPTER 2

Explanation, Possibility, and Determinism

Part of the legacy of the science of organized simplicity is the ideal of a single equation or simple set of equations as the foundation of all our explanations. Indeed this is a form of the reductionist ideal that has the most ardent adherents. The companion ideal is, of course, the explanatory system as deductive system. Although their program was never fulfilled, and to most people today seems unfulfillable, the Logical Positivists will be remembered as those who pursued this deductive ideal to its (absurd) extreme. Much of what I do in this chapter is meant to discredit the deductive ideal and provide elbow room for a less idealistic but more effective evolutionary science.

The deductive ideal is especially vexing to anyone trying to deal with organized complexity, as will be explored in the course of our discussion of complex systems.[1] But since the point of this study is to build an apparatus to deal with complexity, I will not engage in a direct "refutation" of the positivist picture here.[2] Instead, I try to outline an alternative explanatory rubric for the tasks at hand. It is important to remember, though, that the ideal of deductive success will not be the overriding ideal for this alternative. Certainly, the use of all sorts of mathematical techniques guarantees that deductive relationships will be established as a matter of course, but there is no guarantee that a full explanation of any complex phenomenon will constitute its placement in a deductive *system*.

Instead of trying to fulfill ideals, we will be trying to answer questions— namely, those that arise naturally and reasonably in the course of scientific investigation of complex phenomena as parts of evolving biological and social systems. The extent of our success or failure, as I said before, will be the extent to which we can accommodate the solid results of previous scientific practice and extend them to new phenomena of interest. The unity to be sought is straightforward. We want to learn from the sound work our colleagues have done—whatever field they happen to be in—and to eliminate the gaps left in previous attempts to integrate our understanding of complex systems such as human social systems. Stated this way, I would suppose that these ideals could be shared by orthodox sociobiologists. Fortunately, as I spell out this explanatory rubric I can draw on the work of recent theorists—in particular Alan Garfinkel (Garfinkel 1981), whose insights on explanation I exploit as extensively as I can.

Alan Garfinkel's Views on Explanation

One of Garfinkel's core insights is that every explanation is offered relative to a "contrast space" of alternative possibilities.[3] Any of the explanations proffered with respect to the contrast space will answer a well-posed question about the phenomenon in need of explanation. Different well-posed questions will evoke different contrast spaces, hence differently formulated explanations. The initial example Garfinkel uses to illustrate his point is the wonderful anecdote about Willy Sutton, bankrobber become legend, who, when asked by a well-meaning priest why he robbed banks, answered, "Because that's where the money is."

Garfinkel points out that the priest's question is posed with respect to the space of alternatives shown by the more specific question, "Why do you rob banks *rather than* work as a teller or dig ditches or sell aluminum siding or . . . , etc.?", where the list of alternatives is enormous, but not indeterminate. Sutton, on the other hand, answers the question, "Why do you rob banks *rather than* candy stores, town dumps, biology departments, etc.?", where, again, the list is enormous, and this time as indeterminate as the boundaries of our silliness.

Garfinkel labels his view "explanatory relativity" to emphasize his claim that no explanation is independent of some specified contrast space. He later goes on to discuss what we might call the question of relative privilege of explanatory spaces. That is, do we think, as our scientific wisdom increases, that we find special contrast spaces yielding especially important explanations—maybe even *"the real explanation"?* Garfinkel is committed to the view that there is no *one* contrast space yielding explanations of absolute privilege. On the other hand, his view is perfectly compatible with a full range of judgments that for given purposes, e.g. scientific purposes, some contrast spaces are more important than others. That is to say, some well-posed questions are more important or even better than others as we proceed along investigative pathways. On his view, however, it would be impossible to claim that any one investigative pathway was the *only* legitimate pathway in any general sense.

In summary of this part of his view Garfinkel says:

> To summarize, explanations have presuppositions which, among other things, limit drastically the alternatives to the thing being explained. These presuppositions radically affect the success and failure of potential explanations and the interrelation of various explanations. Call this *explanatory relativity.*
>
> A perspicuous way to represent this phenomenon is the device of contrast spaces, or spaces of live alternatives. The structure of these spaces displays some of the presuppositions of a given explanation. (Garfinkel 1981, p. 48)

Garfinkel goes on to show how his conception articulates nicely with one of the most useful and common devices of the sciences, the conception of "state space." This also allows him to introduce a very concrete notion of structural constraint. Since these conceptions will figure strongly in my own views, it is well worth quoting Garfinkel here:

> Basically, these spaces are similar to what physicists call *state spaces*. A state space is a geometric representation of the possibilities of a system; a parametrization of its states, a display of its repertoire. (ibid., p. 40)

and

> ... the imposed structural conditions radically alter the kinds of explanations we give because they constrain and truncate the contrast spaces. There is some precedent for this way of talking, and some good examples are to be found, in the state spaces of physics. (ibid., p. 45)

State spaces can be called "heuristics," "models," "epistemological devices," or whatever you like.[4] But whatever we call them we cannot be justified in imagining a real world "process" of the successive piling up of constraints as *necessarily* being mirrored in our successive manipulations of a matrix. The notion of state space is an analytic tool *we* use to design research. With it we can hold real complexity at bay long enough to allow us cognitive access, within the bounds of our capacity to comprehend complexity. We can analytically manipulate a state space (or a close enough analogue) to determine covariances, threshold effects, and many other interaction phenomena. But we must notice that even with a state space-like heuristic device the design of research is a very complex matter. In general, no one experiment or carefully prepared observation will yield definitive results. The move to a broadened heuristic baseline will require us to seek robust results built from a multiplicity of research pathways. In building these robust results, it is overwhelmingly probable that mere additivity of constraints will fail, so any eventual synthesis will be quite complex (Levins 1966).

The advantage of the state space heuristic is that, in effect, it allows us to relativize constraints to a variety of backgrounds, each thought of, in turn, as the primary causal nexus. Given this heuristic choice of baselines, different aspects of a very complex evolutionary dynamics come to the foreground depending on the particular question being asked. Each question is asked against a background set by a particular line of research.

Anyone familiar with recent controversy in evolutionary biology can probably see how this view of explanation is going to be exploited with respect to evolutionary phenomena. Developmental constraints, stability conditions at the molecular level, thermodynamic constraints, and, potentially, a further host of structured structuring structures (*baupläne* without the metaphysics or the implied statics) will be considered as the material conditions determining the possibility space within which evolutionary events can take place. It may take a good deal of patience to articulate the determinants of the possibility space of, say, a single species; but once you do there is still room for a dynamic of natural selection.[5] Furthermore, questions of determinism within evolution will finally be answerable on the basis of an observation of particular phenomena, rather than being sandbagged *a priori.*

This is enough from Garfinkel to get us started. I will have opportunity to explore the richness of his view further in the course of the book. Since he is primarily (though not exclusively) concerned with a theory of explanation, however, and I am primarily (though not exclusively) concerned with evolutionary dynamics, I deploy his insights in a particular way, and my reformulations will sound a bit different from his original formulations. In the end, however, I do not deviate very much from his line of thought.

Modality, Law, and Totalization

It is important to notice that in setting out the rudiments of Garfinkel's theory the notion of possibility is central. Within the sciences of organized simplicity, modalities such as "possibility" and related notions such as abilities, powers, capacities, and the like recede to the background, or even disappear. The main reason modalities *seem* to be dispensable in the sciences of simplicity is that the equations describing the trajectories of processes (e.g. ballistics phenomena) allow us to depict the past, present, and future in such a way that they can all be "seen" at the same time as a single determinate curve in Cartesian space (or a suitable topological analogue). It is as if we were looking at a total history from a timeless god's-eye point of view. Further, as Prigogine strongly emphasizes, the trajectories considered are in the most important senses time-independent. For our concerns, the question arises whether evolutionary phenomena can be so depicted; as I have said, the view presented here is that they cannot.

Fortunately, the use of the extended state space as a main representational device allows us to say quite determinate things about evolutionary trajectories without making assumptions about time independence, or the pseudo-simultaneous contemplation of entire single-track trajectories. That is, we can move the modal notions into the foreground again,[6] and think of the well-articulated state space as the scene for *possible* trajectories—and, more important from an analytical point of view, a depiction of the ruling out of *impossible* trajectories. I will be advocating the strategy "Look for, discover, and impose systematically more complex constraints." This quite naturally comes down to a search for the limits to what can happen, an articulation of what *cannot* happen. This strategy does not presuppose that there is always a way of finding an ultimate recipe for determining the single possible pathway an object *must* take as its future course. In other words, I do not *presuppose* that all processes amenable to scientific investigation are deterministic. Of course, I also do not rule out the possibility that some processes are deterministic. The effect this has on our ability to predict will be discussed when we look at the concept of determinism.

Just as the economist constructs commodity spaces and production possibility spaces as Cartesian representations of possible future trading or production activity, so we will be able to construct possibility spaces for organisms and ecosystems which depict possible evolutionary trajectories, *given the large array of constraints under which the organisms or ecosystems find themselves.* In addition, it will still be possible to do justice to what we know about the physical "laws" which everything "obeys" (though the language of "law" and "obey" will be scrutinized).

Later in the book, for example, we will be concerned with the way in which the second law of thermodynamics constrains pathways through the possible. It does so by finding a measurable quantity whose magnitude must vary in very particular ways over the course of time. Once we find that this law holds, we have an explanatory resource available for answering certain questions like those of the last paragraph. For instance, when asked, "Why doesn't that perpetual motion machine keep going?" we might reply, under the right circumstances, "It can't. In a closed system entropy can't decrease." Utilizing laws in this way involves putting them in "extremal form" (e.g. Nagel 1961), a technique which is relatively familiar. In fact, this way of looking at physical laws was suggested by Popper many years ago, and has earlier antecedents; but Popper remained tied to the traditional picture of explanation and did not exploit his insight in the way I will.

Consequently, it is my view that anything that can reasonably be called a natural law in the scientific sense simply (a) specifies general constraints, and (b) says that anything under those constraints under appropriate conditions of closure is constrained to a single path, i.e., its trajectory *under the appropriate closure conditions* is completely determined. The "determinism" and "universality" of these laws is usually overplayed by stating them as symmetries and thereby *assuming* the closure conditions (usually equilibrium conditions) under which complete determination is present. For example, students are taught to respect them as laws by encountering them always under conditions where the solutions to equations compare favorably with the "realities" exhibited by laboratory apparatus. "Experiments confirm laws" means nothing more than the conditions of experiment will allow; and what they will allow is law-abiding artifacts.[7]

Now, this notion of natural law conforms well with the practice and *scientific* aspirations of normal science. In fact, it supports a fundamental respect for the scientific enterprise better than competing views of natural law do. For example, when a particular set of constraints becomes well established—e.g. the set of constraints provided by thermodynamics—it becomes established along with the closure conditions circumscribing the set of problems to which it can be applied with the best results. In addition, it establishes boundaries that circumscribe the range of candidates for explanation at any higher level. It defines a sense of reality for us which cuts down the number of fools' errands we are liable to undertake as we try to understand complex phenomena.

We have to be a bit careful, of course. For example, physicists locked into the thermodynamics of closed systems tended for a number of years to impede the understanding of biological phenomena. Physicists' insistence, in fact, on using closed-systems models for biological processes virtually demanded vitalistic speculations on the nature and origin of life. This accounts for the crucial importance the appearance of Schroedinger's *What is Life?* (Schroedinger 1945) had for biologists. The last thing biologists wanted to be stuck with was vitalism (Haraway 1976), yet until Schroedinger (and Prigogine) gave a suitable account of how the thermodynamics of open systems could illuminate the physical constraints on biological systems and provide models for information flow at the molecular level, the alliance of biology and physics remained unsatisfactory.

It seems to me that the transition from the pre-Schroedinger to the post-Schroedinger period is now being mirrored in the development of sociobiology. That is, the reductionist, determinist, closed-system effort to establish the continuity between biological and social phenomena is an honest attempt to eradicate the mysticism from explanations of social systems. But, like pre-Schroedinger thermodynamics, the current sociobiological theories inadvertently impede productive research and understanding. No more than biological systems do social systems "violate the laws of nature," but neither are explanations of social complexes reducible to explanations of phenomena at a lower level of complexity, in the sense that they can be derived from them.

We have to remember that explanations at *any* level relate certain abstracted features of phenomena that have many features. These abstractions define contrast spaces and, within those contrast spaces, are as justified as can reasonably be expected when they answer questions about the behavior of the phenomena in question. They do not have to provide answers to *all* possible questions about the phenomena any philosopher with an overactive imagination might want to ask, nor, at an initial stage, do they have to account for features other than the abstracted features figuring in the answers they do provide.

However, the modesty of this view is intolerable to those whose ambitions for an explanatory theory are to totalize it. So, for example, the Laplacean ambition is to show that *every* answer to *every* question about *every* phenomenon has an answer within Newton's mechanics. Now, *we* know that this ambition can never be fulfilled. So modern day Laplaceans are forced to look elsewhere for a theory to totalize—e.g., to a unified field theory accounting for all the basic (known) forces. Then it would be possible to claim that all *kinesis,* that is, all change of any sort, is to be accounted for by invoking *only* the forces integrated in the unified field theory. In short, the claim would be that *only* those forces can make anything happen, and the presence of *any* feature of the world is to be accounted for by those forces acting on the basic constituents of the world. The constituents are precisely those features that figure intrinsically in the theory. To paraphrase a bit of outmoded biology, ontology recapitulates philology.

This position is metastable for two related reasons. First, as the case of Laplace shows, the choice of a theory to totalize is constrained by the history of the investigative praxis of science. And there is no way out of this. Even if the current best theory gives us all the answers to all the questions we can ask about all the features of the world we can currently discern (which has never been true even of the best theories of any period) we would need further assurance that we had asked all the legitimate questions there are to ask and discerned all the features there are to be discerned. It is not clear where such an assurance could come from. Furthermore, our experience has been that, as science advances, new features of the world are discovered in the very process of trying to account for the old. The examples are ubiquitous: mass, subatomic particles, hydrogen bonds, developmental programs, the dynamics of transposable elements, and so on.

Thus the second cause of metastability arises. A totalizing science makes claims about its own future that cannot be grounded in its past history. The *ultimate* claim—that past progress underwrites a belief in future perfection—is belied by every project of any kind which looked promising in its early stages, but petered out before its ultimate ambitions were realized. An argument that science is somehow special, and can confidently be expected to fulfill its ultimate ambitions, would have to be a nonscientific argument, as yet unavailable, which would, incidentally, defeat the totalizing claims of the science it supported. Meanwhile, however, we can go right on with our sciences as we know them in the absence of totalizing success. Indeed, we can do nothing else.

In the absence of any way of knowing that a science can answer all possible questions, an alternative strategy is commonly adopted. The claim is that the science can answer all possible *legitimate* questions. This simply shifts the problems of totalization up a level to problems of establishing legitimacy. There is no reason to think that the would-be totalizers can solve their problems at the higher level any better than they can at the lower level. We have cautionary tales from the history of attempts to capture criteria of legitimacy—for instance, the history of the rise and fall of logical empiricism—which ought to warn us of the probable fate of attempts to totalize criteria of legitimacy.

As many have pointed out, abandoning the ambition to totalize is fully consistent with affirming the ability to claim universality for basic physical laws. Nothing, absolutely nothing, so far as we know, *disobeys* basic physical laws (universality). On the other hand, it does not follow from this that answers to questions about why something is doing what it is doing must be couched exhaustively in terms of these laws (totalization fails). To that subclass of scientists who are distressed by this rejection of totalization we can only say that their aspirations for omniscience and omnipotence are not underwritten by their own scientific enterprise. We can also point out yet again that their aspirations derive from an earlier, nonscientific, tradition

But, in addition, in the next chapter I will offer an alternative assurance, namely that any feature of the world we can now discern, no matter how esoteric, arose as a consequence of an evolutionary process; and this process began, so far as we know, with the big bang—that is, with the beginning of time and space. The resulting view will be a rejection of synchronic reduction (i.e., a rejection of totalization) and an insistence on diachronic continuity (i.e., a rejection of mystery, vitalism, etc.). I think that the resulting view is stable whereas the totalizing view is not. It can accommodate the historicity of investigative practice and the novelties it generates, and it makes no claims to ultimate perfection.

The Terms of the Contrast

The major differences of the explanatory picture being presented here from the more traditional view of explanation should be becoming clear, and an overview may make them clearer still. The key is to make the primary device for explanation the construction of highly structured contrast spaces. Explanation consists of providing an account of the differentiation of the possibility space in such a way that the future occupation of positions within the space by given objects (in a very broad sense of the word) can be understood systematically. In the differentiation of the space, paths are laid out along which objects proceed. An explanation either accounts for the availability of the path, or accounts for why the object followed one path rather than another. In addition, in many cases which we will be closely concerned with, the *rate* at which things move through possibility space will be crucial.

A science, thought of in terms of this picture, is a systematic account of the ways in which differentiation takes place in the possibility spaces of various sorts of items. It gives a systematic account of the constraints, enablements, and limitations which constitute the pathways through the possible. "Laws," with respect to this picture, are, as I said, treated as constraints to which objects are subject: the constraints that give a particular cast to the differentiation of their possibility spaces.[8]

The possibility space picture is more general than the traditional and more familiar atomistic, mechanistic picture—which becomes a special case of the possibility space picture. In fact, it has been such an important special case that it has appeared to be the general case to much of Western science. And a very great deal has been learned by science operating "exclusively" in the atomistic, mechanistic mode. My strategy is not to delegitimate this mode of explanation, but rather to detotalize it. I want to make available the resources of the broader picture while retaining the local advantages of the narrower. Rosen (op. cit.) makes a similar but more precise point.

Determinism

Determinism appears within the possibility space picture when the constraints on possibility space are so numerous or so strong for some object that there is one and only one path through possibility space available to it. Within the possibility space picture, then, the "determination" language of the more standard view will be accommodated by talking of structuring constraints of various sorts operating at various "levels" or in various "dimensions." For example, a fully adequate account of speciation will require (at least) an account of thermodynamic developmental, and selectionist determinants of evolutionary pathways. It is likely that none of these determinants can be reduced to any of the others. But this remains to be shown.

The possibility space picture allows for the possibility that when *all* the structural constraints have been accounted for a situaton is still not constrained to a single pathway. That is, determination may not yield determinism. In contrast, the atomist, mechanist picture requires that full determination yields determinism. This is why probablistic explanation always has an unsatisfactory second-best feel within that picture.[9] Consequently, the possibility space picture is kinder to many sorts of scientific advance than the deductive-nomological model. For example, it seems to me highly unlikely, given all we know, that human beings will evolve into amoebae. And I think that from current biology I could assemble a series of constraints that would effectively rule out that possibility. But it would be a far different task to show a single path that human evolution *must* take. And I consider a science that can rule out our evolution into amoebae a successful science, despite its inability to predict the evolutionary course of the human species. We are also not going to evolve into buttercups, . . . etc. And think how far we can go along this line without ever being able to establish a determinism. I count this progress and am strangely unmoved by philosophic worries to the contrary.

Scientific activity itself is far more clearly explicable with the possibility space picture, since differentiating (conceptualizing) activity is itself carried out *within* a possibility space—not in a timeless Platonic heaven. There is no avoiding this. It is impossible to be outside possibility space. To think that you are is already to have differentiated some activity (conceptualization) from others; and this is to assign conceptualization its place in possibility space. This is to say, somewhat differently than usual, that no epistemological foundationalism is ever successful and that no god's-eye viewpoint is available to us. Recent philosophy has taught us that in more standard terms. This reinforces the choice of a concrete starting point within scientific practice itself, rather than the foundational starting place favored by post-Cartesian philosophy.

Within the science of biological evolution, the possibility space picture is immediately congenial, allowing us to organize constraints and limitations on possible futures for determinate organisms, species, etc. By ruling out impossibilities (on a whole raft of grounds), we can give a shape to the possibility space an evolving entity actually confronts. For example, think of asking if a particular set of fossil remains is the remains of an evolutionary progenitor of *Homo sapiens.* This naturally resolves into a question of possible (and plausible) evolutionary pathways. We have to ask, on the basis of the best theories we have, what changes would have had to take place for the species represented by the fossils to have evolved into our own species. What range of possibilities was contained in the genome? What mutations were possible and nonlethal? What are the chances that the species encountered an environment that exerted just the sorts of selection pressures which would favor the emergence of human traits? What was the array of developmental constraints?[10]

Now, at the present time we do not know the answers to all those questions for any one set of fossil remains with any great certainty. Some of the questions will, in fact, never be answered. This shows us something important about the determinants of the pathway science may be constrained to take. The state of relative ignorance, some of it permanent, is why the paleontologist finds it essential to build a fossil record with as few gaps as possible. The "dense continuum" fossil record sought by traditional paleontology is meant to provide the evolutionary pathway phenomenologically in the absence of the ability to specify the constraints of the pathway precisely and with theoretical certainty. When theory is lacking, it is tempting to adopt uniformitarianism as heuristic canon.

Astronomers before Newton faced an exactly parallel situation. From a "neutral" starting point, in the absence of a theory that could specify (in this case) deterministic constraints on planetary objects, it was essential to collect minute observations in order to establish the phenomenological continuity of orbits—to fix their pathways. (Of course Tycho's observations were not really carried out in the total absence of an organizing theory, since he had all the theory imbedded in the astrology of the day. In particular, that theory required that planetary motion be uniform. This allowed him to point his telescope in the right direction each night.)

Uniformitarianism in evolutionary theory obviously fills the same role as the theory of uniform motion in pre-Newtonian astronomy. The "dogma" of uniform motion could only be given up when a mechanics sufficiently powerful to supply a successor appeared. The same seems to be true of the uniformitarian dogma. For we need to have sufficient theoretical constraints on possible evolutionary pathways *and* ways of explaining the presence of organisms (and perhaps species) at given points on the pathways if we are to break free of the necessity for super-dense and minute continuities in the fossil record. This is to say that a theory such as punctuated equilibrium could not offer a significant challenge in the absence of a lot of background theory deriving from genetics and developmental biology. Only now that the supporting theory is available, can the new theory begin to be assessed.

Determination and Mechanical Metaphors

Confusion and ambivalence about the nature of determination shows up over and over again in attempts to discern "mechanisms" in situations where processes are better understood in nonmechanical ways. Euphemisms begin to multiply as analogues to force in the classic Newtonian sense are sought. The most common euphemism these days is the verb "to drive." It has been used to describe chemical processes, metabolic processes, and transport phenomena across membranes, the so-called "energy driven," "entropy driven," and "gradient driven" processes. Similarly, in the world of computers, we speak habitually about "disk drives," or "menu driven" software. I have already referred to the "default drive."

I think that the metaphors "drive" and "driven" signal that a reconceptualization is necessary for understanding to move forward. In some (but surely not all) cases, the possibility space picture helps us achieve clarity. For example, in the sense of force or power, what drives my word processing operation is the electric current coming from the wall outlet and the force I expend banging on the keyboard. All the other "factors" that bring about the final product are structuring factors. They organize the possibility space in such a way that the electric current and my typing can combine to produce a desired outcome. Most of the structuring involves the opening and closing of a vast array of switches, producing a determinate configuration through which current flows. *Volkswriter Deluxe* does not drive my computer; it structures its internal configuration. So does DOS and so does the hardware design.

The phrase "entropy driven" is an interesting one. The basic idea expressed by "entropy driven" is that all processes must obey the second law of thermodynamics, which says that in all real processes overall entropy cannot decrease. The law primarily sets accounting limits of a certain sort, limits within which every process must remain. In addition, in the classic cases, the production of entropy can be measured—say by measuring the heat produced by a chemical reaction. But when this accounting system is coupled to a mechanistic picture it often results in an uncontrollable urge to lapse into teleological talk: a lapse which, again, signals conceptual difficulty. For, while people know better than to take the teleological talk perfectly seriously, they still find themselves saying that (say) molecules *seek* a maximum entropic state, or that a system *seeks* a regime of least entropy production.

Everything I know about the concept of entropy—including its formulation in information theory—leads me to think that in discovering the entropy-accounting system physicists have discovered a basic limit to the possibility space of any process. The second law is a classic case of a statement of what *cannot* happen. You *cannot* build a perpetual motion machine; you *cannot* get more energy out of a process than you put in; etc. As a basic constraint on the possible future of any phenomenon, the second law is a tremendously powerful tool (so powerful, in fact, that there is an almost overwhelming tendency to extend it beyond its capacity). Many times, for example, when we want to know how a system will evolve, given certain inputs, we know how to confine its future possible pathways simply because we know that it must obey the second law. The second law tells us that the universe, at least locally, is structured in a way that limits energy flow and material configuration. The impossible configurations can be "read off"; and sometimes, when the system is a suitable one, the remaining possible configurations are such that they allow us to make predictions that future states will fall within a very narrow range. The second law is very seldom powerful enough to introduce enough constraints by itself to allow us to make deterministic predictions. Other constraint considerations must almost always be added to it in order to lead to predictions of unique outcomes.

Later we will examine processes amenable to analysis in terms of nonequilibrium thermodynamics—one of the potentially more fruitful extensions of classical thermodynamics. When we do so, I will try to avoid talk of "entropy drive" and instead talk of systems responding to the structuring of thermodynamic limits, attempting yet again to vindicate a move from the atomist, mechanist picture to the possibility space picture.

In addition, any definition of basic forces depends on establishing an "inertial" baseline, the state of no change. Many familiar explanatory theories, not just Newtonian mechanics, set up an inertial baseline. For example, classical Darwinism maintains that evolutionary events will not occur unless the force of natural selection (or some other "biasing agent" such as genetic drift) acts to "cause change" (Sober 1984). Theories that base their explanatory structure on equilibrium conditions also establish a sort of inertial baseline. The pattern emerges that baseline events need no special explanation, and deviations do. So, for Aristotle, the state of rest for terrestrial things and the state of circular motion for celestial things need no special explanation; for Newton, uniform rectilinear motion needs no special explanation; and, for the neo-Darwinist, the maintenance of Hardy-Weinberg ratios needs no special explanation. In each case, the justification of the inertial baseline, and the scope within which the justification will be successful, has to be questioned. If, for example, the relevant ambience is not static but in flux, then the inertial baseline will be totally misleading. An entity that is apparently in equilibrium may in fact be undergoing dynamic change.

What is obvious in the cases of Newtonian forces and selection forces is slightly less obvious in the dynamics of biological and social life. Especially in the latter case, we ought to be very careful about the establishment of inertial baselines.[11]

The key situation to be reconsidered in this regard, as an illustration of the way the possibility space picture can work, is the sociobiologists' invocation of the default drive discussed in the introductory chapter. This invocation is a natural consequence of a commitment to the traditional explanatory picture—helped along by a bit of liberal ideology. The traditional explanatory picture requires individuated "forces" operating in a suitably defined inertial framework. Complicated forces are vector resultants, decomposable pseudo-orthogonally into simple forces. Liberal ideology provides the premiss that within societies the only "actors" are individuals—acting within an essentially inert nature. Apparent social complexities are vector resultants composed of individuals' actions. At best, these individual actions result in social vectors of an optimal sort—as if guided by an invisible hand.

On this picture, deductive-nomological models seem to be the main vehicles of social explanation. Individual actions constitute the antecedent conditions of a law whose consequent is some social eventuality. Or, to put it another way, every social explanation is couched in terms of the solution to a decision problem. Either the decision is actually the *rational* decision of an agent or agents, or the agent is assumed to be the locus of instincts or innate responses which, over evolutionary time, have become keenly adapted to the production of appropriate behavior parallel to that which would have been rationally chosen. When a result does not constitute an optimal solution of the decision problem of an individual, and when no "coordination problem" can be identified which is solved by a given result, then the only alternative is to think of the "result" as the (unfortunate?) vector resultant of more or less incompatible lines of rational action as basic forces.

Of course, *given the premisses,* this may not be the silliest way to conceptualize matters. For the theory is a theory of agency from the start; and the candidates for nonpersonal agency are often pretty bizarre. But explanation in terms of the attribution of agency is itself a pretty confining picture of explanation. For one thing, it is an inertial theory that never specifies its principle of inertia. The consequence is that, given this picture of explanation, biological determinants can only be thought of as quasi-voluntaristic, hence, by default, bestial and assignable to the default drive. This is a good example of the way in which ideology constrains explanation: that is, how it guides science down a particular narrow explanation pathway.[12]

Material

The next chapter provides a rubric for treating issues associated with "reduction-ism," "physicalism," and "materialism," as they arise in an examination of the evolutionary dynamics of complex systems. The position I argue for is already prefigured in the main concepts developed in this chapter: possibility space, structured structuring structures, determination. One more concept needs to be discussed in preparation for the lines of thought pursued throughout the remainder of this book: the concept of "material." In particular, the phrase "material condition" needs to be understood clearly as I intend it.

For many people explicitly concerned with theology and the metaphysical foundations of science, "material" has settled into a very specific conceptual slot, contrasted with "spiritual," "ideal," "mental," or some other dualistic opposite. "Material" is then roughly equivalent to "physical" (though finer distinctions are possible), and we can suppose that the physical is what the current generation of physicists certify as the real and basic stuff of the universe. If "material" is read this way in what follows, puzzlement is sure to arise.

Using "material" as a contrast to "spiritual" has well-known drawbacks. It virtually locks us into the small set of Cartesian alternatives: we can be dualists of some sort[13] or pursue a reductive path and become "materialists." For "material" in such a contrast space tends to prioritize a conception of *what is* in terms of a particular theoretical framework (usually contemporary physical theory), then tries to totalize that theory as the unique source of contrast spaces for fully legitimate or fully scientific questions and answers. But each such conception requires a process of abstraction to move us from an ordinary sense of what is to the specialized sense articulated in terms of the favored theory. Yet, while such abstractions are an absolutely essential condition for systematic sciences, and are justified insofar as these sciences provide answers to important questions we ask, it is quite another matter to turn these abstractions into metaphysical theories. No one has ever provided a compelling argument for doing so, except for the purposes of theological speculation. Such speculation is none of my concern. Furthermore, it has been traditionally argued (most recently by classical positivism) that a unification of our scientific activity requires establishing a prioritized reductive base. But it will be part of the argument of this book that our attempts at unification should be more sophisticated.

The usage of "material" I adopt in preference to the one just discussed is a perfectly ordinary one, in no way "extended" or unusual. I would say, in fact, that the metaphysical acceptance of "material" is the unusual one, and far more historically bound than is usually noticed.

Certainly we are all familiar with building material, writing materials, the material our editors want us to cut from our manuscripts, and the material from which our clothes are made. If we travel by air enough we may even be familiar with smoking materials. This familiar sense of "material" is the core of the concept as it appears throughout this book. The space within which this sense of "material" functions is the space of making, fabricating, constructing—which is precisely why I adopt it. For the dynamics of evolution is a dynamics of construction: the shaping and reshaping of available materials. And while "constructing" hints at a "constructor" we are surely sophisticated enough to discount any such anthropomorphic baggage when it is inappropriate.

The link with the notion of structured structuring structures is quite obvious, since such structures are the organized and organizing material out of which subsequent organized and organizing material arises. The sorts of questions the concept of material helps to answer are questions about the conditions of possibility for various phenomena (in the next chapter, the phenomenon of self-replication, for example). So the specification of the material conditions for this and that will be a constant task. An appropriately constituted, appropriately organized Haldanian soup, for instance, is, on most views, a material condition for the origin of prebiotic and biotic systems. But, equally, one of the material conditions for the space shuttle program is our ability to solve complicated ballistics problems using the calculus. Those finding themselves quite able to accept Haldane's soup as a material condition, and unable to accept our ability to use the calculus as a material condition, have a fine opportunity to reexamine their metaphysical reflexes. In particular, a reflex dualism may have come to light. As I said, I have nothing to contribute to metaphysical disputes, and my use of "material" should allow those disputes to go on in the specialized usage they depend on. There needn't be any necessary connection between my sense of "material" and the specialized use. In particular, I see no reason to think that decisions about the latter have to be made before we develop a theoretical framework to deal with evolutionary dynamics in terms of the former.

My use of "material" has a further advantage in working out accounts of evolutionary processes—again, because of another very usual sense of "material." "Material" is customarily contrasted with "immaterial," especially in law, to mean something like "relevant" and/or "important" as opposed to "irrelevant" and/or "trivial." This is convenient, because every phenomenon occurs in a busy world where lots of things are going on, and we surely want to say that some things occurring before and along with phenomena we are interested in are relevant, and others are not. The phrase "material conditions" will always intend this sense.[14]

Finally, we should notice that for something to be material in the full sense it has to be "the right stuff." Plastic wrap is not normally material for pants. Culture and sensibility might change so that plastic wrap becomes acceptable for pants material, or, in a pinch, we might *have* to use plastic wrap for pants, but this simply reemphasizes the historical contingency of material. In a similar vein, trees are not often building material these days, though they once were. However, trees are still very often *potential* building material: the material from which building material is made.[15]

We can not go on to see how all these materials for an explanatory framework contribute to an understanding of the evolutionary complexification of the prebiotic and biotic worlds.

CHAPTER 3

The Structure of Emergence

My overall strategy in this chapter is to show that emergent phenomena occur at the molecular level; examine certain important features of emergence; and go on to suggest that much of what is true of emergence at the molecular level is true of phenomena more commonly said to be emergent. This last step is essential for a diachronic materialism capable of dealing with society and culture.[1]

In conformity with the stance adopted in the previous chapters, I again insist that the ultimate success of a project seeking to establish continuity of investigative method throughout the range of phenomena from molecules to societies cannot be decided on the basis of *a priori* considerations. It depends on the continuing success of research programs which expand their compass stepwise into new areas. I will argue throughout that any synchronic reductionist program will fail in this. But I will also argue that there is no reason to think that a diachronic materialist program will fail. Indeed, unless you believe in miracles, it cannot fail. The sociobiologists' conviction that society must be understood in continuity with biology is correct. Their instinct that the continuity must be established on an atomistic reductionist basis is mistaken. The conviction of some philosophers that cultural phenomena are irreducibly intensional is correct. Their instinct that this is inconsistent with materialism is mistaken.

To give an overall sense of the enterprise, I can state some of the conclusions that I think can be established. First, each of the evolutionary stages I will sketch (and any alternatives to them) involves the emergence of structured structuring structures.[2] Second, the presence of these structures allows processes to occur in ways they previously could not; or it gives rise to rates and determinacies that could not previously have been achieved. Third, processes that occur as a (partial) result of these structures cannot be explained without specific reference to the structures; such reference is ineradicable. That is, once structures emerge, synchronous reduction to more "basic" phenomena is impossible.

But, fourth, diachronic explanation of these phenomena is always possible. That is, the natural history of the preparation, development, and emergence of the structures must always be given in terms of earlier phenomena. These earlier phenomena themselves have a natural history. They too emerged, and eventually the natural history of all phenomena is a matter entirely in the realm of physics, in the sense that at some time in the past no structures existed which were not objects whose behavior could be accounted for exclusively in terms of what we currently recognize as physical theory.[3]

The General Problem

My strategy is to show that emergence occurs at the level of molecular evolution, let alone at the level of human culture and society where it is usually noticed. I will start with an observation by Herbert Simon (Simon 1973) that continues to be the most concise foundation for evolutionary thinking. Simon points out that if the probability of the existence of complex plants and animals is considered as the probability that (say) a random mix of atoms will assemble themselves into (say) an aardvark, then the probability of evolution occurring is as close as you please to zero. But, says Simon, this is a total misconception of the evolutionary process and the assessment of its probability. For the evolutionary process is modular (lumpy). From an initial starting point the probability of the evolution of simple stable modules is significantly high, so it is no surprise that these modules appear. Given the existence of the *stable* simple modules, the evolution of stable higher modular complexes of the simple modules is significantly high (as, of course, is the rise of transitory, unstable ones which will prove evolutionary failures), so the appearance of the higher modules does not strain credibility at all: self-assembly of the appropriate sort is known to occur under reasonable conditions. And given the availability of successive accounts of the processes which give rise to successively more complex modules (now to be considered as hierarchical systems in Simon's view), the appearance of the aardvark is relatively as probable as any other natural contingency. On the Simon view it already follows that synchronic reduction of (say) aardvarks to atoms is impossible. For no view that simply analyzes aardvarks as concatenations of atoms can account for the discrepancy between the probability of aardvarks on its view and the probability of aardvarks on the (Simonian) evolutionary view.

One thing to notice here is that a commitment to irreducible emergent structures in no way entails a commitment to supraphysical entities which violate physical laws. Emergent structures are perfectly compatible with materialism of a reasonable sort. Every emergent structure bears the marks of its genesis. The atomic events going on in and around the aardvark are obeying (say) the laws of relativistic quantum mechanics. Furthermore, at some earlier stage on the evolutionary trail that led to aardvarks, events primarily describable in quantum mechanical terms are decisive in "driving" the processes which resulted in early structures. That is, in the light of the discussion of the last chapter, quantum mechanical considerations by themselves determined the possibility space of the processes. As the structures became more complex, events at the atomic and sub-atomic levels became constraints and limitations on further evolutionary events, but not necessarily decisive for their production. That is what it means to say that the emergence of a structure alters the probability of succeeding events.

This is precisely the sort of situation that requires us to reexamine the agenda of science. The search for universal laws was thought to be a paramount activity for science. Or, failing global universality, laws were sought which held perfectly generally within well-defined domains. A more promising view is that the advance of physics involved not simply the discovery of laws, but the discovery of packages consisting of laws and the boundary and closure conditions under which the laws held. The boundary and closure conditions are not derivable from the laws, but define the conditions for the existence of solutions to the equations in which the laws are stated.[4]

Within the domains defined by the appropriate boundary and closure conditions, processes governed by the laws are subject to *smooth* transitions. That is, holonomy and integrability hold. But this ideal is the wrong one for dealing with the evolutionary phenomena involving emergence, for such evolutionary processes are *lumpy*. They are understood not by extrapolating smooth holonomic accounts to new domains, but by understanding the modifications of boundary and closure conditions in a way that makes holonomy impossible. H. H. Pattee has seen this more clearly than most theorists and has argued that the main problems confronting those who want to understand the emergence of living systems are problems of hierarchical interfaces.[5]

My account is meant to begin to do justice to Pattee's insistence that we approach hierarchical interfaces from both sides. My claim, following Bhaskar, is that this can be done only by rejecting synchronic reduction and accepting diachronic materialism. Once stated, however, problems of accounting for the relevant hierarchical interfaces remain (as the cognoscenti like to put it) "highly non-trivial." Each interface presents a combination of conceptual and empirical problems that have to be dealt with simultaneously. They have to be treated philosophically, but cannot be treated within philosophy abstracted from first-order scientific practice. Call this a higher order interface, if you like. Philosophers in general, and physicists in general, have been at loggerheads over problems of hierarchical interface because (as Pattee points out about the disagreement between Crick and Michael Polanyi) each approaches the interface from one side only. *As* physicists (exclusively) or *as* philosophers (exclusively) they can do nothing else.

Emerging among the Molecules

I will begin by sketching the account of molecular evolution that Manfred Eigen and his coworkers give.[6] This is one of several accounts of molecular evolution that are currently alive. Certain aspects of it will most certainly find their way into any eventually acceptable account as investigation in the field continues. Other aspects will be supplanted, and, in any case, the view will be supplemented.[7] For my purposes, however, the eventual fate of the theory is not of crucial importance, for, in important ways, the Eigen account is a "worst case scenario," and all theories which now contend with it are far more congenial to the points I want to make than it is. The details of this situation will emerge as I proceed.

The theory begins with the obligatory Haldane soup, redolent with organic molecules. Within this vast array of molecules are some which subsequently formed the constituents of life on earth, and many others which no longer figure in the molecular economy of living things. The problem set for the theory is to account for the ascendancy of the correct molecules rather than the incorrect ones, and then for the rise of recognizably living forms.

The first order of business is to account for the synthesis of molecules at significantly differential rates so that the composition of the soup could evolve in directions more favorable to biological correctness. This story must eventually involve an analysis of differential stereochemical affinities, selectivities, tacticities, stabilities, etc. For the worst case scenario presented here, the primary concern is the emergence of suitable replicating macromolecules. (Later we will briefly examine alternative accounts involving phase separated systems, but for now we can continue to focus on the replicators.) Despite the (apparent) primacy of DNA in the genetic information chain in later stages of biological evolution, persuasive arguments are given in favor of RNA as the initial significant polynucleotide replicator. The chief arguments concern the differential capacities of DNA and RNA for chemical activity in the absence of complex, finely tuned enzymatic systems found only in highly evolved cells. DNA seems not to be active enough to account for the relatively rapid evolution of replicating systems which must have occurred. As soon as DNA strands complex enough to be significant replicators occur, they coil themselves up, curtailing their replication (let alone transcription) activity.

This leads to the consideration of various strands of RNA floating in the soup. Some have "a homogeneous stereochemistry and with the correct covalent bonding in the backbone of the strand could reproducibly lead to stable secondary structures" (Eigen et al. 1981, p. 91). Such strands have two advantages. Their particular three-dimensional structure makes them more reliably stable than other strands, and the same three-dimensional structure contributes to their capacity for relatively high fidelity replication (hifi rep). So, over the course of time their tribe will increase, while no other identifiable sort will become significantly more numerous. Stability *plus* generation of equally stable replicas must go together to account for the increased prevalence of the "correct" sort of RNA. Equally important to all replication, of both "correct" and "incorrect" molecules, is an energy source, and, complementarily, the replicators' ability to utilize the available energy. Many studies have demonstrated the likelihood of natural conditions in which energy sources utilizable for RNA polymerization occur.

Experiment has shown that "the mechanisms of template-induced synthesis [of RNA strands] and of template-free synthesis are quite different" (ibid., p. 96). In particular, the rate determining step in template-free synthesis is stochastic, whereas in template-induced synthesis it is deterministic. Now, the context of the investigation which produced this result was quite specific, the synthesis of RNA in the presence of Q_{beta} replicase. But there is no reason to believe that other circumstances are logically very different. What the results show is that the emergence of template-enzyme systems gives rise to a discontinuity in replication rate, and this discontinuity is a consequence of a change in the mechanics (at least in the rate-determining step) of replication. This means that an account of replication within the template-enzyme system requires referring to the presence of the system. Conversely, replication at certain rates becomes evidence that certain systems are present. More generally, the rate of template-induced synthesis must be accounted for by the presence of the template, an emergent structure. On the other hand, the *fact* of template-free synthesis (and, indeed, enzyme-independent synthesis) allows us to infer that the emergent template structure need not be the result of a more basic chemical process. In other words, we have scientific warrant to accept the emergence of such structured structuring structures as templates.

Next we can consider a sequence of events and conditions which affects both the rate at which "correct" molecules accumulate and the conditions for their continued evolution. First, any enzymatic activity prompting the hifi rep of the "correct" RNA will certainly increase its advantage *vis-à-vis* other molecules. In fact, the situation may arise in which the concentration of enzyme in the vicinity of the replicator becomes rate determining instead of the intrinsic replication capacity of the replicator itself. (Of course, if phase separation occurs as physical or chemical boundaries arise creating reliable "insides" and "outsides," then stability and replication success will be greatly affected, but here in our worst case scenario we put this aside.)

Eventually *hypercycles* can develop. In the simplest cases these consist of coupled, mutually catalyzing processes such that the rate of increase in each is greater than it would be (or could be) if uncoupled. For example, suppose that replicator A codes for a protein which serves as an enzymatic replication enhancer for replicator B, and replicator B codes for an enzyme enhancing the replication of replicator A. Then, in an environment in which a myriad of replicators and proteins were competing for available precursors, the coupled system of A and B would have a significant (Darwinian) advantage. In fact, computer simulation and experiment show that under many circumstances the A-B system could drive out all competing replicators.

The theory of hypercycles seems to me one of the most exciting developments in evolutionary theory over the last few years. Willingness to emphasize the centrality of hypercycles presupposes a willingness to think in terms of emergent systems, which was rare in earlier evolutionary theorizing. In this case, the elegance of the mathematics is a particularly clear and persuasive justification for thinking of evolutionary states as structured structuring structures, for the mathematics straightforwardly exhibits (a) the structural conditions which lead a prior system to the unit simplex of the hypercycle; (b) the structure of competitive interactions as the hypercycle organizes the distribution of its components (often through recognizable selection events); and (c) the evolutionary restructuring consequent upon the action of the hypercycle (especially catalytic hypercycles). For example, we see readily how "In particular, a catalytic hypercycle, once selected, will not tolerate (in homogeneous solution) the nucleation and evolution of independent competitors. The genetic code and the chirality of systems which have emerged from a single hypercycle are therefore universal" (Kuppers 1983, p. 209). An additional advantage is that for those with limited mathematical background the structure of the system can be set out fairly rigorously by using directed graphs. This potentially ties the analysis of hypercycles structurally to the techniques of ecological analysis set out by Levins (1968), and so potentially offers quantitative techniques to supplement the essentially qualitative Levinsian analyses.

Hypercycles may also be important in integrating some of the results and conceptions of the strand of molecular evolutionary thinking represented by Fox and his students, thus alleviating the tension between this line of thought and that of Crick and his followers. For example, the following is a clear invitation to analysis in terms of hypercycles as structured structuring structures.

> Crick (1968) focused his attention on the "origin of the genetic code." He considered two explanations for that origin: (a) stereochemistry and (b) a "frozen accident." Since it has been possible to demonstrate selective recognitions between polynucleotides and thermal polyamino acids experimentally . . . , it has been further possible to infer that the basis for the genetic code is rooted in the stereochemical forces, or reactivities and shapes, of molecules. The determination of dominance of nucleotide by (a) content of lysine in copolyamino acid and (b) the reflexive determination of content of lysine by identity of nucleotide in polynucleotides (Fox 1974) suggests that these two complementary recognitions would have been locked into an evolutionary feedback process. Such a process would, then, have the quality of a "frozen" event. The processes that were frozen were fixed on the core of interactions between amino acids and nucleotides, each of which was part of a more complex set of substances and mechanisms.
>
> These experiments thus lead us to the view that an important primitive development was frozen stereochemistry, to paraphrase Crick. But the freezing was no accident; the nonrandom interaction of molecules of different shapes . . . was the predisposing cause rather than an accidental occurrence (Fox 1975). (Fox and Dose 1977, p. 245)

Recent work has both deepened this picture and made it more complex, so that the role, say, of hypercycles *per se* may be a good deal more modest than our worst case scenario suggests. The general point stands, however.

Finally, to leap over many additional evolutionary steps, suppose we have a system consisting of coupled RNA replicators, furnished with an efficient enzyme system and a reliably present boundary package capable of selective permeation conducive to stability, hifi rep, and transcription. Now we can suppose that deoxyribonucleotides are floating in the environment and can travel across the boundaries. They could well be a thermodynamic burden to the system. That is, from the point of view of the internal economy of what are now *very* highly evolved phase-separated microsystems (protocells or true cells, depending on what is counted a cell), the deoxyribonucleotides make no contribution to the efficiency of the microsystem, and, quite possibly, decrease its efficiency. Thus, initially, the microsystems containing these "fellow travelers" could even have a lower probability of survival than those without them. But suppose that through the normal mutation process an RNA strand coding for a reverse transcriptase (an enzyme promoting RNA-induced DNA synthesis) appeared in the system. This may seem suppositional, and is. The point is that it is far less implausible than competing scenarios. The result would be the synthesis of DNA strands, initially nonfunctional. Since the RNA templates available for DNA synthesis would naturally be the primary functional ones in the system, the nonfunctional fellow travelers would in effect become secondary information bearers. Without a method of transcription, however, this information record would itself be a thermodynamic burden to the system. Under conditions of thermodynamic closure, i.e., at the boundary of stability,[8] systems which had the DNA burden would be selected against. The systems need not, however, be under thermodynamic selection pressure. Furthermore, in some systems the replication machinery already present might provide the initial conditions for an evolutionary process in which the superior hifi rep potential of DNA could swing the probability of survival in favor of microsystems that had DNA information storage. As is known, this would be extremely unlikely in the absence of complex enzymatic systems, methods of proofreading and repair.

Should such an evolution take place, then the resulting environment would contain both DNA and RNA replicating systems. The complex apparatus favoring hifi rep of DNA and the complex apparatus of DNA transcription are assumed to have evolved. Under such conditions the DNA replicators would have an enormous advantage. As opposed to RNA systems, for example, they would have the capacity to store long messages—coding not only for extremely large structural proteins, but also for a complex regulatory system—and the plasticity to support cell differentiation and integration.

We know that DNA systems are presently predominant, and that the few RNA systems that exist are dependent on the presence of DNA systems. In fact, DNA systems are predominant enough to have underwritten the "central dogma of genetics." We can note, though, that if the account in the last two paragraphs is *possible* on the basis of current knowledge, then the central dogma and its attendant linear management hierarchy fall short of being established. (Of course this failure of the central dogma is not as contentious or challenging as it would have been a few years ago. Challenges to the central dogma seem to be coming from all directions.)

It is clear that the story of the evolution of DNA information storage which I have sketched is schematic, incomplete, and speculative, but not fantasy. It is consistent with what is now known and is conservative in that it seems to be a part of all currently viable accounts. It is, however, still biased toward the "master molecule" view of replicating systems. This bias is deliberate, since it allows us to examine the related bias toward linear building block accounts of evolutionary processes, as opposed to multiple systems feeding positively and negatively into one another.

In the full flush of research following the solving of DNA's structure and the working out of the genetic code, the dogma of linearity of information transmission was posited as a guiding fixed point of genetic research. This dogma quickly eroded, it is true, but the bias of linearity has not, despite a number of considerations that undermine it. Leaving aside biochemical considerations for a moment, let's just note that *even if* all biological information pathways at the present time were DNA-RNA-protein pathways (and of course they are not all of that sort) it still would not follow that the evolutionary sequence of events in the genesis of biological systems was the evolution of DNA, evolution of RNA, evolution of protein. Yet on the linear building-block model any other sequence would be well nigh impossible, since only DNA seems an informationally powerful enough building-block to support the subsequent edifice. That is, the linear theory is committed to an analogue of the medieval dictum that the reality of an effect can only be as great as the reality of its cause—namely, that the "first cause" in the linear sequence must contain in itself all the information necessary to complete the sequence. Since the only constituent of biological systems that can contain such information is DNA (by reason of critical length, stability, replication fidelity, etc.), DNA is identified as the first cause.[9]

DNA *systems* are the predominant biological systems at the present time, and this predominance needs to be explained. But, of course, it does not follow from this predominance that reproductive and epigenetic processes proceed with DNA as the active mastermind (as opposed, say, to a filing system), or that DNA must be predominant in any story of molecular evolution. Furthermore, if DNA systems emerged out of earlier systems, then their very evolutionary success would obscure their origins from the investigative field of the linear building-block theorists. Fortunately, the field of molecular evolution seems to be on the way to overcoming this danger—precisely by developing nonlinear models of great promise. In my view, the same bias toward linear models infects thinking about emergent systems such as societies. In those cases, too, models of systems evolution must be developed which can overcome the bias. The next important thing to learn from the emergence by DNA systems is that they had to emerge *as systems*. DNA cannot replicate, nor is its replication fidelity superior to that of RNA unless it is coupled with a complex enzymatic system of proofreading, repair, and replication promotion. These systems almost certainly were prepared by pre-DNA systems which put together the conditions upon which the final evolution of DNA systems rested. The evolution by DNA itself of the replication system it now enjoys is an unlikely if not impossible scenario.

A social phenomenon to compare with the emergence of DNA systems may be the rise of market society. When in place (or when imagined to be in place) the market can seem to be a powerful self-standing foundation of social order. It can seem as if all other social institutions may be subordinated to the market as secondary devices to facilitate its operation. But perhaps a closer look at the generative dynamics of market systems would show the complex of pre-existing subsystems which had to be in place before the market system could evolve, and indeed must continue to be in place if the market system is to be a tolerable component of social order. This may be true *even if* in certain competitive circumstances market systems can drive out alternative sources of social order and become virtually ubiquitous.

I introduce this example here only as a potentially helpful illustration. But it is an illustration of an important sort. For one of the critical issues in our time is that of reductionistic and "holistic" methodologies in the social sciences. Uppermost in current intellectual debate are the questions of reduction raised by sociobiology, but there are other longstanding disputes about methodological individualism in the social sciences yet to be resolved satisfactorily. Worth mentioning are the failure to integrate micro and macro economics, the distinction between the internal and external understanding of another culture, and the distinction between voluntaristic and institutional explanation. In addition, there is a now classic example in the philosophic literature which focuses the issues at stake here quite well.

We are asked to imagine all the "parts" of a university—buildings, professors, students, etc.—and then to concede that the university is no more than the sum of these parts. But, taken together, these parts may or may not be a university. Now, organized one way, with a particular evolutionary history both with respect to internal organization and with respect to a society at large, the "parts" are parts of a university. Given some other organization of them, some other evolutionary history, at some other time, they need not constitute a university at all. The university is contingently emergent (as are all emergent entities) and could as contingently disappear *without* the death of any people and without the destruction of any buildings.

When you try to state the difference between a pile of building materials and a house, you contribute to understanding by saying how the materials are reorganized. Reorganized, they constitute a new set of possibilities, equally, a new set of limitations—both consequences of the way the materials were put together. Put together another way—other possibilities, other limitations. This case is no different, except in the generative mechanisms, than the case of atoms in cells. Now, what could make the university ontologically different from the house or the cell? What could make it less real? Nothing, as far as I can see, for I can give an account of it as a structured structuring structure, and answer questions about its genesis that take the mystery out of its existence. On the other hand, if prevented from referring directly to structure, organization, systematically arranged constraints, and evolutionary genesis, I would not be able to make sense of the university at all. This seems to me to be true of all social and cultural entities, just as it is true of self-replicating molecular systems.

Analysis and Synthesis

Emergence and reduction are deeply entwined with the concepts of analysis and synthesis, so we need to discuss how they fit with the present view. It is somewhat difficult because the exact nature of analysis in general is still contested, largely because any process of analysis is bound to a context of techniques and fundamental assumptions about the constituent nature of the world. But chemical analysis is a successful, well-established practice and can serve as a basis for discussion.

Qualitative and quantitative chemical analyses are without doubt an important part of the investigation of living systems. So much is platitudinous. But no amount of precision and thoroughness in determining the atomic constituents of a biological system can yield an explanation of how the system works, let alone why. This is true no matter at what level the analysis is carried out. Fortunately, a *wealth* of techniques is available, and explanations that could never be built up out of analyses at a single level are available by creatively combining techniques of many levels. The multiplicity of techniques and the theories that tie them together make the cross-inferencing that yields robust explanations possible.

One of the best examples of such success (though there are innumerable examples) is the building up of the picture of primary, secondary, and tertiary structure in proteins. Combinations of techniques were devised which allowed chemists to solve the atomic structure of single peptides, the linear sequence of peptides, and, finally, the three-dimensional shape of the protein which, for example, accounts for its enzymatic activity or structural role. Those who like a good intellectually satisfying detective story would enjoy retracing the investigative path of discovery. Without going into all the technical details of how protein structure is solved, we can still make an important point about analysis, synthesis, and reduction.

Since tertiary structure is a consequence of linear sequence, and linear sequence is (at least stochastically) a consequence of primary structure, and primary structure is (at least stochastically) a consequence of the combinative capacities of individual atoms, a reductive pathway seems to be generated that confirms the synchronic reductive view. In fact, however, the string of consequences confirms my own view. For we have to look at the *way* in which each higher order structure is a consequence of lower order structure. To do so, we can contrast two scenarios.

First, on the basis of careful chemical analysis we determine the empirical atomic composition of a protein (say one evolutionary variant of cytochrome C). That information, coupled with the selection of a suitable quantity to work with, will tell us how many of each atom we have to put into a container to provide sufficient building blocks for a given quantity of cytochrome C. (Notice that the scenario is already absurd.) We put all those atoms into the container and wait for the laws of physics to do their job and give us a container full of cytochrome C. Now in this case (analogous to earlier cases we have examined) the probability of ending up with a full container of cytochrome C is essentially zero. The chemical analysis yielding the atomic constituents of cytochrome C gives us the atomic bookkeeping, but the bookkeeping, even coupled with the laws of physics, fails to explain either the presence of the protein in our world, or why the protein has the properties it does.

The alternative scenario is essentially evolutionary. In effect we could try to re-evolve cytochrome C in the laboratory. Of course we needn't try to reproduce exactly the evolutionary steps toward the actual emergence of cytochrome C in organisms. But any synthesis must avail itself of the emergent structures which make the synthesis possible. In the laboratory this would undoubtedly involve a careful choice of precursor molecules and catalyst, purification processes at various stages of synthesis, etc. Starting from (atomic) scratch is ruled out by time considerations, which is not simply an inconvenience for synthetic purism, but rather a concession to the necessarily evolutionary character of synthesis.

An apparently trivial, but, in fact, paradigmatic illustration of this point is the following. You can imagine deciding to buy your kid a bicycle for Christmas. You look in a catalogue and see one that seems just right at a good price. At the bottom of the entry it says "SOME ASSEMBLY REQUIRED," but you judge that you're up to putting a bike together, and send for it. A few weeks later a truck rolls up and delivers a large basket of iron ore, a large bag of limestone, a large bag of charcoal, a box of bauxite, a drum of sap from a rubber tree, and a lot of other stuff, some of which you can't even identify. The bill of lading says "one bicycle." Synthesis of most things, even bicycles, is ineradicably path dependent on the order of steps, each of which is rate determined. The synthesis requires completion through modular intermediaries. Without structural intermediaries the probability of achieving a successful synthesis is virtually nil and the rate of synthesis prohibitively slow.

This discussion shows something that ought to have been obvious. "Analysis" and "synthesis" can (but need not) be symmetrical. This is partly because both can refer to temporal processes and both (but especially "analysis") can refer to atemporal logical reconstructions.[10] The history of the two concepts is, of course, fascinating. (For example, Hegel can be read as claiming that analysis and synthesis are always symmetrical and always refer simultaneously to historical processes and rational reconstruction.) In general, all combinations of conceptions of analysis and synthesis have been put forward at one time or another, but the most important point for my purposes is that the conceptions often get mixed in such a way that unsound inferences are permitted. In particular, atemporal analysis is commonly put forward as correlative to temporal synthesis. This is the usual move in arguments in favor of atomistic synchronic reduction.

But here the difference between the *bookkeeping*[11] of the constituents of a complex phenomenon and an explanation of the phenomenon is clearest. To show that some complex phenomenon is reducible to a set of more "basic" phenomena, it does not suffice to show that the sum of the "basic" phenomena constitutes a complete bookkeeping account in terms of some inventory previously chosen. It is necessary to show that an explanatory pathway can be traced from the basic phenomena to the complex. If the pathway requires ineradicable reference to the temporal emergence of intermediate structures, then, as I have said, the synchronic reduction fails even though the diachronic explanation succeeds. Thus analysis in terms of bookkeeping atomism is fully compatible with the ineradicable existence of structures not represented in the basic bookkeeping inventory.

It must be remembered that for any complex phenomenon there are many possible paths that an analysis can take, each yielding a basic bookkeeping. Each of the bookkeeping inventories will be connected with explanation in a different way. The consequence of this is familiar, and has already figured in my argument. Any *single* analysis can (and usually does) fail to provide an explanation of evolutionary genesis. This is yet another asymmetry between analysis and synthesis, an asymmetry which is often obscured by the adoption of Newtonian analytic simplification as the operative paradigm.

The fundamental claim being contested in this section is that analyses are, or could be, path independent, dehistoricizable techniques. The multiplicity of chemical analytic pathways, the necessity of combining the results of several of these pathways, and the complexity of experimental inference involved in successfully using the pathways all argue for the path dependence of chemical analysis. The sophistication with which analytical techniques in chemistry have been developed and utilized, in fact, has given them a degree of autonomous legitimacy virtually tantamount to the path independence some scientists dream of achieving. It is only when we notice how chemical analysis threads its way through the maze of chemical structures, exploiting structural features of the chemical environment and basing its experimental inferences on its successive discovery of structure, that we can appreciate its dependence on the historical pathway of discovery. And, even then, knowing the path dependence of chemical analysis is always of minor importance, and sometimes of no importance at all—thanks to the overwhelming successful praxis of chemistry. Only when extravagant adventitious reductionist claims are made do we even gain the right to worry about the status of chemical analysis as an investigative technique.

Identity and Commensurability

Radical worries about identity arise from two sources, it seems to me. The first is familiar—the urge to totalize the entire investigative enterprise in terms of a single super-practice such as the Unity of Science program. Such a program would indeed need an overarching univocal *criterion* of identity rather than an array of strategies for establishing identity in and across interacting investigative practices. The second source is the tradition of looking at "bodies of knowledge" as self-contained languages, and languages on the model of closed formal systems. In this framework, problems of identity between two bodies of knowledge have to be thought of as translation problems, or problems of establishing formal equivalences. It is no surprise that on this view the problem of radical incommensurability between two investigative traditions, *conceived as alternative formal systems,* has arisen. Along with it has arisen a literature on the establishment of "interfield" bridges, principles, links, and translations. But the problem itself is an artifact of conceptualizing investigative practice on the model of a formal system.

A few years ago it would have taken a much longer, sober, argument to deal with this "problem." For then the dream of establishing well bounded self-contained "fields" was still alive among serious persons. "Sciences" were dreamed of as dehistoricized, axiomatized (or axiomatizable) quasi-formal systems, with their semantic content gained through foundational evidence of immediate sensation—*protocol Satze,* etc. But (and talk about difficulties with identity criteria!) no way was ever found to demarcate *either* the putative axiom systems *or* the foundational evidence. For the theologians of simplicity and one-dimensionality these failures were a crushing disappointment. For those whose interest was getting on with the job of understanding the natural world, the failures were of little consequence. But you can see how crucial it would have been for the dreamers of formal systems to "solve" the "problem" of the commensurability of alternative access paths. They required the definitive sealing off of paths, one from another, for their formalization to succeed. And so, in effect, they introduced the very incommensurability they then agonized over.

Most (but not all) sophisticated investigative practices certainly do involve the use of language. And sometimes strategies such as axiomatization—as an act of housekeeping—are useful. Sometimes things need to be tidied up in order to keep track of where you are and where you might go next. But the tidying up—in the context of a changing, complexifying array of investigative practices—is highly provisional, and has to be done again and again in any investigative practice that has not degenerated into dogma and stagnation. When sciences are thought of as activities organized into practices with well-established critical dimensions, not only does the need for rigid demarcation between them disappear, but such rigid demarcation promotes fragmentation—an impediment to integrative advance and the possibility of mutual aid. A sharp demarcation between activities and practices is a feature of a life of institutionalized schizophrenia—familiar enough in a world requiring, say, a clear distinction between self and bureaucratic persona—which, in my experience, does not seem to be the life of the practicing scientist. When scientists in various subdisciplines learn from one another, and they do, they confute the dream of science as a set of nicely bounded formal systems. Issues of autonomy then reduce to jurisdictional disputes in the bureaucracy of science, and very few working scientists have time to worry about them much except at faculty meetings.

As far as the issue of incommensurability is concerned, I do not have much to add to what Ian Hacking (Hacking 1983) has said. Beyond that, the only sensible thing to do is to go see how identifications are established in present scientific practice and examine the history of the sciences for instructive instances. As far as present science is concerned, there is no scarcity of interesting cases of the successful establishment of identity. Every issue of *Science* or *Nature* provides new ones. Latour and Woolgar (1979/1986) are particularly good at showing how identities were established (constructed) in the case of releasing factors. Similarly, careful work in the history of science is equally fruitful in showing how important identities are established. In fact, the more familiar you become with scientific work itself, the more embarrassed you feel about raising questions of radical incommensurability.

At this point we need to recall—as a cautionary tale—what genuine incommensurability would be. The dispute between evolutionists and creationists is again our most convenient case in point. Honest versions of the two theories are genuinely incompatible and, in terms of the ontologies they support, incommensurable. The consequence—the existential consequence—of trying to conduct a life informed by both theories, a life within the discourse of science *and* the discourse of literalist Christianity, is fragmentation, systematic self-deception, bad faith, and all those other impediments to "authenticity" the old existentialists used to warn us about. In other words, the incommensurability between scientific and religious discourse *can* be a truly serious one. Against this background, the attempt to look at the differences in approach and agenda between, say, the biochemist and the molecular biologist as incommensurable is simply frivolous.

We can concede that temporary terminological problems will arise to slow down the rate at which they learn from one another, and lament the fact. But this is an issue of far different dimensions than that between evolutionists and creationists. The terms of the sharing of discourse are decisively different.

Ontology in Practice

Questions of identity within the context of social research are somewhat more complicated than the equivalent questions within, say, biological research. Indeed, a central part of Foucault's research project is to trace the ways in which human beings have been constituted as objects for social scientific investigation. Questions of reflexivity must be considered carefully, for, in obvious ways, in social research the "objects" investigated and the "subjects" investigating them are the same people. Yet what the natural sciences seem to teach us is that questions of identity are tractable if they are handled intelligently within the confines of well-designed research; as a first strategy, we might well begin to deal with questions of social identity in those terms.

Within a well-established transactional praxis you begin to notice a "phenomenon." That is, something keeps happening to you, to your experiments, to your investigational field, that is pervasive and stable enough to become a potential "object of investigation" in its own right. Now, eventually what you want to mean by "object of investigation" is something that happens independent of your investigative access to it, but which you *can* get investigative access to. But an object of investigation in this sense is something that has to be won, not something that can be presupposed.

So the primary cognitive field is a partially reflexive investigative matrix: an object *cum* investigative means *cum* investigator system which has to be examined dialectically from *inside*. What you are trying to do is determine the internal structure of this complex system. In the end, you hope that this internal structure can be managed so that a case can be made (and nailed down) to the effect that investigator and means can be sealed off from an object in such a way that the object can be treated as an independent system *looked at from outside by the investigator*.

Different phenomena will be differently tractable to such a sealing off. Quantum mechanical phenomena offer special problems (Schroedinger's cat, etc.), as do anthropological problems (participant observer, etc.). Newtonian mechanics has been spectacularly successful in abstracting a set of objects that can be detached from observers: so successful, in fact, that it has perpetuated the illusion that we approach phenomena from outside *ab initio,* that external access is a starting point. But in any case the Newtonian dream is a powerful one. It is only when we notice (1) how limited its success really is, and (2) that we can reformulate *it too* in terms of winning access to sealed off observation that we can break its hold long enough to give ourselves a chance to reproduce its success in areas that do not offer its *ease* of success.

The process of differentiating and determining the investigative space with a view to detaching the object is bound to vary with respect to the structure of the praxis within which the process begins. But there is one feature which seems to me common to all such processes. They depend on the development of multiple pathways of access. In the simplest cases this is achieved in the repeatability of experiments, or, in the wild, public access to data (anybody can go look at the arctic lakes). But multiplicity of access does not stop there. Intersecting techniques have to be developed for us to be sure that what we consider to be an object is not just an artifact of a particular technique. An example of the difficulty of being limited to one or a narrow range of techniques can currently be found in high energy physics where "particles" keep popping up as potential independent objects, and their status is subject to considerable question. Quite similarly, "minds" have always been a similar source of difficulty because of asymmetry of access and the ideological freight of individualism. As we saw, areas such as protein chemistry offer a fine object lesson in the way that robust results can be obtained at the intersection of a plurality of sophisticated investigative techniques.

In chapter 5 we will be concerned with the nature of members of a hierarchy. Problems of identity may well arise there, for members of hierarchies are at the same time identifiable in many other ways. Only rarely does membership in a hierarchy "exhaust" a person's identity. So in the light of what I have said here let's recognize at the outset that for good and sufficient reasons individual human beings are readily identifiable to other human beings. But identified how? Are we to think that a dehistoricized account can be given for some primary identification? No, there is no such dehistoricized account of the identification of humans. Yet, for any proposed identification, another can be provided. We always have multiple access to one another. This is a consequence of all human life being polysystematic.

Life being what it is, the most common identificatory scheme we have for others is visual appearance: we describe one another in terms of height, hair color, etc. Notice that this is far from a Newtonian framework, although we say when we give such a description that it is physical.

Social life being what it is, we also very commonly identify ourselves by name, residence, and family. Residence is spatial, though the identification of that residence has more to do with Rand McNally than Newton, but, as identifiers, name, residence, and family are thoroughly social. They are also very often the decisive identifications. Italians almost always refer to popes by family name, for example, because family (hence residential) origin is an important political link to the legitimacy of the church hierarchy at a particular time.

Finally, as a definitive "physical description" we have come to use finger prints—and these days voice prints and even blood types for identifications of a particularly decisive kind for particularly sensitive purposes. It gets harder and harder to be an imposter.

CHAPTER 4

Complexity and Closure

Implicit in the preceding chapters has been the claim that the standard neo-Darwinian explanations are inadequate to cope with the complexities of evolution. Here I want to make that claim explicit by examining one of the favored techniques of selectionist explanation, the theory of games. I will do so by focusing on the ways in which such explanations have to be circumscribed in order to be acceptable.[1] Explanatory closure is a much more difficult achievement than is sometimes thought. When we have examined some of the major difficulties in achieving explanatory closure, we will have a much more realistic sense of the usefulness of game theory and, from that, a much more realistic sense of the usefulness and limitations of selectionist explanations.

Closure

The general form of a closure condition is shown by the following: Assume that A and B are the alternatives. Either A or B; not A, therefore B. The closure condition is the assumption, "A and B are the only alternatives." While certainly not always as simple as the example suggests, closure conditions are ubiquitous in our explanatory activity, and are often so well understood that they go without saying. When we do want to make them explicit, we can often do so rather easily. We say "what we're talking about," that is, we specify a domain of discourse in a way that pins matters down in an acceptable way. Or, if we are engaging in formal experiments, we design them with an appropriate system of controls which serve to establish closure. Or, when we have well-established and well-worked-out scientific theories at our disposal, we can often derive closure conditions from them. Often boundary conditions yielding closure are consequences of the properties of particular mathematical formulations. A major part of the Galileo-Newton-Descartes achievement of establishing a mathematical physics consisted of identifying a small closed set of physical parameters sufficient to provide explanations. The solar system ramains as the outstanding example of a natural physical system which is closed with respect to its mechanics (though it clearly is not closed from, say, an information-theoretical point of view). In some areas, such as orthodox thermodynamics, specifying boundary conditions and other closure conditions is a constant task. The standard physical symmetries (conservation laws) require precisely defined closure.[2]

Science, with physics as its *de facto* model and experiment as its primary investigative technique, has achieved undeniable successes. But the adoption of physics as a norm has meant that problems of systems closure have been treated in a relatively standard way. From a mathematical point of view, system closure has been assumed to hold for a set of phenomena when a mathematical model of the phenomena could be produced that was formally similar to a model that held for a closed physical system. When a genuine question of closure is raised (in terms of the adequacy of the model), an independent argument must be found to establish closure. But, more usually, closure is posited as a satisfactory approximation, *ceteris paribus*. This latter tactic is nearly universal in the construction of, say, econometric models, but it also occurs frequently in population biology. Rosen (1986) makes a similar point about Ashby's mechanistic assumptions.

Closure conditions quite often appear to be "dimensional," as when it is said that the solar system is mechanically closed and thermodynamically open. This dimensionality could disappear only if all the "dimensions" could be reduced to one. The Laplacian program was, among other things, an attempt to perform just such a reduction in Newtonian terms. Philosophers are familiar with problems of dimensional closure in issue after issue generated by the Galileo-Newton-Descartes program: Primary and secondary qualities; the mind/body problem; the relationship between reasons and causes in human action; the establishment of identity *simpliciter* as opposed to "identity under a given description." I think these are all problems involving closure and boundary conditions in complex interactive systems, and are best studied as such. Here they may at least serve as reference point analogues.

Questions of closure exactly parallel to the issues just mentioned arise in evolutionary biology. In its progress over the last century biology has dealt with some of the questions amazingly successfully. As in physics, some problems of explanatory closure have lent themselves to solutions in terms of experimental design or in terms of theories specifying the range of operative variables. However, robust explanations within evolutionary biology have proved more difficult to obtain than was hoped. In my view, a good part of the reason the difficulties have arisen is that the complexity of evolutionary (and ecological) systems has not always been sufficiently appreciated. In attempting to achieve explanatory closure in terms of simple, one-dimensional mechanical systems, orthodox neo-Darwinians are cut off from ever fully understanding the phenomena they study. This self-limitation was indeed for a long time necessary and productive, but as the program has tried to extend itself to issues in molecular evolution and sociobiology, the limitations have begun to look particularly serious.

An examination of game theoretical models will serve to exhibit some of my worries about explanatory closure in evolutionary explanations. I think that there are interesting uses for game theory within evolutionary biology, but there are also major pitfalls.

Bookkeeping and Teleology

It is important to notice that game theory in its standard applications is a bookkeeping theory, and not, by itself, an explanatory theory. For instance, suppose you watch a game of tic-tac-toe and see both players pursue an optimal strategy. The game ends in a draw. You ask, why did the game end in a draw? The answer you get is because both players pursued optimal strategies. This may or may not be an explanation, depending on what you have packed into "pursued." There are innumerable reasons why someone playing tic-tac-toe might execute a sequence of moves which constituted optimal strategy from a bookkeeping point of view. (Learners do it by accident sometimes.) In this stupefyingly simple example, to be sure, there is hardly any room for alternative explanations. The natural immediate inference from bookkeeping to explanation is forced upon us, since there are virtually no plausible alternative explanations available. However, the easy inference is an artifact of the trivial example, especially since it is an example of a sequence of actions that is utterly meaningless if not performed in a game, as a game, and for the sake of winning the game. The general meaninglessness of the activity provides closure, yielding the inference to the one explanation that remains as a plausible candidate. More complicated activities must be treated with more sophistication. But in each case, any inference from game theoretical bookkeeping to explanation requires the imposition of closure conditions.

For games in the literal sense, the usual conditions producing explanatory closure are that the players are playing to win; are under no external constraints; and are "rational" in the appropriate sense. These closure conditions added to the bookkeeping yield an explanation[3]—though, as is often pointed out, they tend to yield tautology if care is not taken to leave open the possibility that, say, the players could have chosen not to play to win.

A second sort of closure that is required to extract explanations from game theoretical bookkeeping is what we could call payoff closure. We know that to be able to model a phenomenon as a formal game, we must be able to construct (or at least provide a recursive recipe for) a payoff matrix, and we have to be able to make sense out of the entries in the matrix. For example, one of the most important measures used in, say, sociobiology is Hamilton's notion of inclusive fitness (Hamilton 1964). This measure defines relative success and failure in the evolutionary game in terms of numbers of descendant gene bearers. If this measure is accepted, then it appears that there is at least a hope of specifying payoffs for useful game theoretical models. But there is serious doubt about whether the measure should be accepted. It follows immediately from its simplistic use that *E. coli* are more successful than human beings. But perhaps this is an unfair extension of the concept of inclusive fitness, and it ought to be confined to intraspecific comparisons, whereupon Catholics win the evolutionary game when they play against Protestants. We could, by brute force of will, decide that this was what evolutionary success meant—adopting thereby the rather odd teleological criterion of numerousness. We have to note, however, that bacteria, at least, have achieved their success while sacrificing complexity (and quite possibly by *avoiding* complexity). This presents us with an interesting choice, for some people wish to construct their evolutionary teleology in terms of higher and lower levels of complexity, with humans at the pinnacle of evolutionary success, despite the fact that they are less numerous than, say, earthworms.

Now, adopting an evolutionary teleology will provide closure of the payoff matrix. Indeed, game theory works for games in the literal sense because the teleology of game winning can normally be assumed. For evolutionary phenomena, the assumption of the teleology of numerousness simply reflects a decision to have the entries in the game matrix represent numbers of descendant gene sharers. If in search of more sophistication you produce a more complicated formula for constructing the matrix, taking into account carrying capacity will not change the nature of the teleological closure condition. Presupposing the teleology of complexity, on the other hand, would require finding a measure of complexity and filling in the matrix on its basis. Several quite persuasive and perhaps useful measures of complexity have already been devised (e.g., information density) (Gatlin 1972; Brooks and Wiley 1986).

There are several reasons for being careful about the difference between intraspecific and interspecific comparisons. Darwin's own theory was concerned almost entirely with intraspecific comparisons. The variation he presupposed was variation within a single species, and, while we must remember that "speciation events" occur, differential survival is almost always differential survival of members of the same species. Darwin's focus, determined, no doubt, by his concern with the "species problem" as it was historically posed, is perfectly understandable. It has to be made clear, however, that the selection of that focus is itself a closure assumption, sealing off interspecific interaction as a secondary consideration. This closure was extremely unstable. We only have to think of how quickly mimicry obtruded itself as a subject for Darwinian explanation to see that it was. There may still be problems that yield to examination of solely intraspecific comparisons, but certainly no genuinely ecological problem does. Insofar as evolutionary ecology has shifted to the core of evolutionary biology, the original Darwinian program has had to be modified. I contend that the shift in models that accompanied the problem shift can profitably be looked at in terms of the various closure conditions that are required to extract explanations from the various models.

The concepts of fitness and adaption are important here. A great deal has been written about them lately. I will simply fit them into the present discussion. All fitness measures, including inclusive fitness, behave formally like budget possibility surfaces in economics. Differences in fitness are exactly analogous to differences in budget. Just as any set of resources in economics can constitute budgetary assets only within a particular set of economic arrangements, exchange system, or the like, so fitness represents assets as they are defined in a particular environment. Just as in economics nothing, not even money, is a system-independent asset, so no trait necessarily contributes to fitness, especially at the margin. However, of course, in both spheres some things are more centrally and surely assets than others. They contribute to success in virtually every economic or ecological environment.

Closure conditions enter the consideration of fitness in expectable ways. Fitness is always fitness with respect to an environment or range of environments. Variation is evolutionary relevant only when it contributes to the fitness "budget" relative to predetermined environmental variables. In principle there is no problem in specifying these conditions in the abstract. The trouble comes in matching the chosen model and the selected closure conditions to the situation being modeled. In other words the number of possible evolutionarily relevant traits and relevant environmental conditions is very large. It takes a great deal of analysis even to get to the point of choosing an appropriate model. Any analysis in terms of fitness must deal with complexity and closure with great care, whether or not the model used is a game theoretical model. The assumption that each organism is either in an adaptive equilibrium with its environment or is pursuing an evolutionary path toward adaptation is one of the most common devices for closing ecological systems for purposes of explanation. Recent literature has amply shown how easy it is for the concept of adaptation to fall into tautology or teleology (Brandon 1978, 1985; Burian 1983; Mills and Beatty 1979).

Most evolutionary biologists, of course, talk as if they want to avoid teleological explanatory closure. Historically, post-Darwinian biology has prided itself on eliminating teleology from evolutionary theory. So it is surprising that subdisciplines such as sociobiology, professing to be Darwinian in spirit, rely on closure conditions such as inclusive fitness which are openly teleological. The best arguments against doing so are their own. The main one is that there is no convincing scientific way to establish one end rather than another as the "correct" or "real" one. At any rate, numerousness is hardly one of the attractive candidates if you are seeking a *telos*. In fact, it is so unlikely a *telos* that it fosters the illusion that it is not a *telos* at all.

Is there a nonteleological way to achieve payoff closure in such a way that we will find game theory useful? We might try the one proposed by Slobodkin and others (Slobodkin 1968, 1961/1980; Tuomi and Haukoija 1979; Tuomi et al. 1985). Slobodkin speaks of the existential game, the idea being that the only "payoffs" are survival and failure to survive. Survival needn't be thought of teleologically any more than the persistence of transience of physical object. Some water is ice, and some is not. Some, bound up in an arctic glacier, will be ice for a long time; some only until the mint julep is gone. Neither quantity of water is a particularly successful piece of ice, nor is either to be disparaged for its failure. The persistence of a crystal and the persistence of an organism can be thought of in the same nonteleological terms. The dinosaurs were not evolutionary failures. There just aren't any around anymore.

This secularized version of assigning game payoffs is promising in a number of ways. First of all, it relates easily to the basic sort of "negative selection" view favored by Darwinians. Second, it allows us to think of the principle of natural selection as the idiosyncratic articulation of one among many concatenations of stability conditions for natural objects at a certain level of complexity. Comparable stability analyses can be provided for atomic nuclei, crystals, chemical compounds, and so on up through a rich array of structures including social structures. The principle of natural selection stands out, and appears qualitatively different, only as an historical consequence of its genesis and of our special fascination with organisms including ourselves. And third, the existential game gives us a sensible way to deal with numerousness, complexity, longevity, and other important features of evolutionary phenomena. We can say that they constitute various *strategies* for achieving payoffs—or, if we're really serious about eliminating the teleology—we can say that they are explanatorily related to the relative persistence of various species, and go on to spell out the explanations.

The basic facts of the evolutionary game are that there are some organisms around; some that were once around but are no longer around; and (more than likely) some that have not appeared yet, but will be around later. The matters to be explained by evolutionary theory are (a) why are some organisms around and not others, and (b) how do new organisms arrive on the scene? From the point of view of the existential game, the question is whether we can hope to explain who the survivors are likely to be, and why they are likely to survive.

The Players

If we are going to talk about "who" survives to play on we are going to have to consider the players in the evolutionary game. The most likely candidates are individual organisms. Individual organisms live and die; and breeding takes place between individual organisms. So, classically, individual organisms are thought to be the players. There are other candidates. Dawkins puts forward DNA as the player(s) (Dawkins 1976). From a different perspective, Lila Gatlin (1972) also suggests that DNA molecules are the players, though her theory differs in crucial respects from that of Dawkins. Others have proposed groups, species, clades, organized independent entities such as hills of ants, etc. Eventually I want to say: "All of the above plus others." But that point has to be arrived at slowly through another examination of closure conditions, We need to get closure on a statement that S, Y, Z, et al. are the evolutionary players, and *they are the only players.*

When game theoretical explanations are closed teleologically, the players are quite naturally identified by locating the teleological agents: The players are the entities who make moves in order to win. The trouble is, though, that with the possible exception of some hominids, none of the potential players in the existential evolutionary game cares whether it wins or not. In fact, precious few of the potential players have the slightest idea that they are even playing the game. Pretending to the contrary does make the use of game models much easier (by introducing teleological closure), but at the same time it vitiates the attempt. The fairytale parallel to a theory needn't shed much light on the quality of the theory itself. This is yet another reason for rejecting teleological closure. So closure on the list of players has to be found elsewhere.

First, let's imagine a typical ecological complex and think about the things we could count. Unless some reason can be found for ruling them out (i.e., an explicit closure condition) each will remain as a candidate for the list of players.

(a) There are species that can be counted—within limits. If speciation events are taking place within the complex, indeterminacies will arise. But the fact is that counting species (on the basis of identified individuals, of course) is one of the first steps in any census.

(b) *Higher taxa* can also be counted. There may not be many occasions to consider them as players, but investigations of macro-evolution could require that we model, say, the competition between angiosperms and gymnosperms.

(c) It is ambiguous whether the *population* of each species at one time is the same as the population at an earlier or later time. "Population" is too flexible a term without further specification. Do reproduction and death constitute a function which maps one population onto another population quasi-continuously? Or do they constitute a function which tracks changes in the same population? The term is used in both ways. Suitable identity criteria can be established to justify either usage. The precise audit in any given case awaits the adoption of one or the other of the sets of identity conditions. If, for example, we want to say that the population of a certain species has increased, we need to adopt one set of identity conditions. If we want to say that a parent population had managed to give rise to a filial population, we will need the other. Lots of people seem to want to say both at the same time. This affects the way in which we might want to define *lineages* as the primary players of the evolutionary game.

(d) *Individual organisms* can be counted. Our ability to do so again depends on some very sophisticated theoretical judgments resulting in identity conditions. For example, a stand of trees where all the individual trunks have grown from one continuous rootstock presents an interesting problem. In addition, our ability to count individuals of a given kind depends upon the theoretical taxonomic assumptions establishing the kind. Counting the dinosaurs which survive in the present day may require a joint effort involving (at least) paleontologists, evolutionary geneticists, and, depending upon the decision reached by those two groups, the Audubon Society and Frank Perdue. This points up the degree of care that would have to be exercised if lineages were the chosen players, for boundary decisions would have to be made respecting both the fact of descent, and the fact of discontinuity sufficient to lead us to talk of speciation.

(e) *Breeding pairs* can be counted. In sexually reproducing species, especially in small populations, this can be an exceptionally important account to keep. But then, in species whose populations exhibit social structure there may well be crucial accounting units (herds, flocks. schools, etc.) of varying sizes which, for example, determine breeding statistics. Do individual antelopes play games against predators, or do the herds play team games?

We now move to a list of the potential players least likely to be candidates for teleological closure.

(f) DNA, selfish or not, has been cast as a player in the evolutionary game. The problem is to establish identity conditions for it that will yield a useful accounting. The situation is intriguing from a philosophical point of view because a consideration of DNA in this context plunges us immediately into a consideration of types and tokens. I think that right at the outset it is possible to reject individual pieces (tokens) of DNA as the "players" in any sort of game. This is especially true when molecular "games" are spelled out in thermodynamic or information theoretical terms. For the information differentials crucial to differential survival are at least as likely to be embodied in stochastic processes as in deterministic ones. Coupled with the fact that individual tokens of any DNA type differ as the cells they occur in are differentiated, this means that DNA eventually acts evolutionarily only through an internally structured complex, making the individual pieces of DNA more like the pieces in the chess game than the players. If it is remarked that the same can be said of individual organisms *vis-à-vis* breeding populations, I can but nod my head and note that both situations need to be worked out clearly in a more ample context. Here I will say only that a consideration of the difference between DNA as type and DNA as token is ultimately crucial to getting a firm grip on the difference between bookkeeping and explanation. It is clear that as people have taken the concept of a genetic language seriously over the last twenty years, they have uncovered problems about the nature of primary "information bearers" which are identical to problems faced in more traditional philosophies of language (Margolis 1983).

Hence, instead of "naked" DNA, (g) *polynucleotide/protein complexes* could be players in an evolutionary game. For it is known that such molecular complexes can bear information and engage in causal process which "naked" DNA cannot (Kaufmann 1985).

Let me end the list of potential players here, though I am far from certain that I have included every important candidate. We have more than enough complexity in view already as we examine the questions of explanatory closure which concern us.

Choosing the Players

We now have to emphasize the systems-dependence of the identity conditions for players as players and a given statistical array as a payoff matrix of a game played by those players. They must be identified together. Atomists will quickly point out that the players, at least, are independently identifiable since, for one thing, we were able to identify individuals and populations long before we started thinking in terms of evolutionary games. This misses the point. First, an example: there are species of insects of which the males and the females were long classified in two different species (and even in two different higher taxa). Caught *in flagrante* they were blushingly reclassified. Similar mistakes have been made with the juvenile and adult forms of the same insect. An identification is an answer to the question "What is it?" Every identification is system dependent. In the case of the insects just cited, breeding interactions and life cycles had to be known in order to arrive at satisfactory identification—even taxonomically. The system dependence of identification is often ignored or denied, since the sloppy, complex megasystem which grows up within everyday life does not seem systematic enough—or is so polysystematic that it seems asystematic. "What is it?" usually gets answered in the routine conduct of everyday life, so that the identification of things seems to be a recognition of what they are in some absolute system-independent sense. Yet any identification is an assignment of place in a (more or less coherent) order of things (Foucault 1970) and is an historical artifact. There is no way that identity can be detached from identification except in ethereal philosophical abstractions. Every identification has its theoretical underpinnings in the evolving epistemological and ontological nexus of life—a life with practical, usually theological, and possibly scientific dimensions. Here we can recall the discussion of identity from the last chapter.

This impinges on our discussion of game theory in the following way: To say that players and the game must be identified together is to say that the systematic constraints of the game model establish a "local ontology" specifiable only within the full account of the game. For *some* games the entities that turn out to be the players will already have been identified in some systematic way. This is no accident (it is a tribute to the ongoing success and expansion of our investigative praxis), but it is also no necessity (new "things"—e.g., galaxies, codons, quarks—may be discovered in the course of our investigations). Our discovery of game models may lead to the discovery of players. It would be sheer dogma to think that the entities involved in biological evolution (i.e., the units of selection) *must* be precisely those biological entities previously fixed upon as the primary biological units. Nevertheless, the continuity of investigative praxis requires the integration of any new projected units into the solid core of an already highly successful biological research program. New units cannot be fadged up arbitrarily and have any legitimate claim to be taken seriously by working biologists. Burian's discussion of the gene is a good illustration of this point (Burian 1985). This is why my list of potential players is confined to the strong candidates I can find on the basis of the literature of contemporary biology itself.

Thus the situation's complexity makes the straightforward identification of games and players more difficult than the literature would suggest. We can see this most easily in a brief analogy. Baseball players are not, from a game-theoretical point of view, the players in the game of baseball, where the payoffs are pennants and world championships. The players are Cardinals, Orioles, Blue Jays, Tigers, Cubs, Angels, and so on—i.e., teams. Now the ballplayers themselves have a special claim to be *thought of* as players, because they are the ones who hit, throw, and field. Just so, individual organisms have a special place in the evolutionary game, because they are the ones who are born, forage, graze, prey, procreate, and die. But the ballplayers are imbedded in a rigid structured structuring structure which radically constraints their activity *as players,* dictates their role in the game being played, and prevents them from being players of that game in the technical sense. The same is true of individual organisms or lineages.

From another perspective, ballplayers are sent onto the field or placed on the bench, given orders where to stand, when to swing the bat, when to run, etc. Since the manager manipulates them, from this perspective the ballplayers are more like the chess pieces than they are like chess players. Like the chess pieces their roles are differentiated, giving an internal structure to the team. You can't win with twenty-five pitchers on the roster, for example. Or, if you have a fast centerfielder, you can get away with a slow-footed, rock-handed slugger in left field, otherwise not. Thus, the selection of ballplayers for a team is seriously constrained by the structure of the game, and, in the last analysis, by the abilities and specialties not only of other ballplayers on the team but of ballplayers on competing teams.

Every ecological environment is multistructured too. So, any analysis of it in game theoretical terms has to rest on a correct account of the structures. In fact, the baseball analogy is much too simplistic a model for natural environments. A closer analogy would be putting together a roster to compete in baseball, basketball, football, and soccer in the same season.

The Games Evolution Plays

There is no reason to believe that a one-dimensional analysis focusing on individual organisms as the sole "players" will succeed. Such an analysis is a highly abstracted ideal type. Pursuing it demands imposing an *a priori* structure that only fragmentarily reflects the goings-on in the natural environment. The proliferation of *ceteris paribus* clauses that occurs when defenders of this ideal type are confronted with structural complexity is the tipoff to the ideal type's inadequacy. In the end, the adoption of the ideal type predetermines an arbitrary choice of closure and boundary conditions which impedes rather than furthers research.

It is easy to see, though, why the ideal type of the individualist competitive game persists so strongly. It seems to offer a clean set of mathematical techniques to deal with the competition assumed to be the main organizing fact of nature. But the mathematics of game theory is only incidentally connected with competition. It merely organizes the differential outcomes of joint activities in matrix form, and examines the consequences of activities which would occasion different paths through the matrix. These activities could be competitive, cooperative, or neither. For instance, in the classic statement of the Prisoners' Dilemma, the players are not initially competitors, but, rather, self-interested and either indifferent or slightly sympathetic to one another. The prosecutor changes the structure of their relationship. The Prisoners' Dilemma is instructive because it shows how structure can be imposed on two people so that they become, from a bookkeeping point of view, competitors. And, if there is an external agency powerful enough to keep that structure in place, the two players may well have to come to recognize each other as competitors and act accordingly, thus converting bookkeeping dialectically to explanation.

Obviously, climatic changes, migration, and most other ecologically relevant events can have the same effect. I see no reason why emergent social structures cannot do the same. On the other hand, evolutionarily important events could have the consequence of ending the competitive relationship between "players," rendering them oblivious to one another, or even landing them in positive mutuality. If, when, and where this occurs is clearly a matter for concrete research, not *a priori* theorizing. Any use of game theory will require analyses at least sophisticated enough to bring all these structures and structured structurings to light.

Sometimes a relatively isolated, relatively self-contained game will be found, especially in appropriate time frames in relatively stable environments. The search for such isolation and independence is, after all, very like the search for system integrity within a hierarchy of near-decomposable systems (Simon 1981). In the usual way, horizontal and vertical uncoupling may occur so as to leave a particular subsystem relatively autonomous for a significant length of time. So, for example, a particular predator/prey interaction may end up being, within an ecological complex of systems, the subsystem decisive for understanding the fate of the members of the system, especially the predators and prey themselves. The genomes of the two may be stable (the rate of change tiny compared to the rate at which they affect each other in the predation relation), the climate constant, the primary food source relatively constant, etc.[4]

This sort of simple closure is exemplified by Darwin's own treatment of genetic variability. His theory posits inheritable, randomly generated variability. This is to say, "Look for no *explanations* of evolutionary events within the genome; attribute them to chance." Randomness is posited in order to achieve explanatory closure with respect to high frequency events, and, at the other end, uniformitarianism is posited to achieve closure with respect to low frequency events. In short, Darwin's theory assumes that middle frequency evolutionary events are sealed off vertically. This allows him to achieve explanatory closure at the phenotypic level.

The language of "high, medium, and low frequency" is also borrowed from Simon (1981). The idea is that in any hierarchical system of systems there are events occurring at high frequencies in the lower-level systems which can be lumped together, averaged, or otherwise summarized in a single parameter (or "sufficient parameter," see Levins 1968) for modeling the middle-level system. Furthermore, there are events occurring at low frequencies in the higher-level systems which are infrequent enough to be treated as occasional exogenous perturbations of the middle-level system. The concept is obviously derived from the investigation of physical systems, but it has persuasive familiar analogues. In astronomy, the mechanics of the solar system can be described without referring to events at the atomic level (high frequency events) and without referring to events of interstellar evolution (low frequency events). In meteorology, weather systems can be described without reference to the motion of air molecules and without reference to secular climatic changes. The analogous Darwinian picture would be high frequency events at the molecular and chromosomal level and low frequency events at the geological and climatological levels.

Clearly, the subsequent synthesis with Mendelian (and Hardy-Weinberg) statistics depends on the same sort of sealing-off assumption. Mendel's laws of segregation and independent assortment are, in fact, closure conditions. They can be questioned quite reasonably, as they have been by Beatty (1981), who points out that segregation and assortment are the consequences of meiotic events which themsleves may be (in fact must be) under genetic control. Consequently, neither molecular and chromosomal events affecting linkage crossing over, nor other obstacles to Mendelian ratios, can be sealed off in any simple way ir order to allow explanation at one level only. Thus the move from the Mendelian bookkeeping to explanation is far less straightforward than the standard synthesis would have it.

There is no doubt that the early orthodox Darwinians were right in imposing middle frequency closure. Ignorance of molecular genetics demanded it. Within its limits the program was enormously successful. The anomalous deviations from the standard ratios had to be allowed to pile up as the orthodox program was pushed as far as it could go. However, the orthodox theory is no longer faced with a heap of unsystematic anomalies. Molecular genetics, immunology, information theory, and other fields have cut away much of the ignorance which justified middle range closure *vis-á-vis* high frequency events. To a lesser extent, closure *vis-á-vis* low frequency events has been challenged now that the uniformitarian hypothesis has a reasonable competitor in the theory of punctuated equilibrium. So the old closure assumptions now appear simplistic and arbitrary.

From another angle, the simplicity and arbitrariness of the orthodox Darwinian program is most glaring when the interpenetration of biology and society begins to be examined seriously. Since the neo-Darwinian pattern requires confining explanations to a single systems level, and is geared up to handle only one-dimensional selection processes, it has no adequate way to deal with the emergence of new social and cultural systems with an evolutionary dynamics of their own.

Reproduction is a biological phenomenon even when it occurs among humans; and differential rates of reproduction continue to constitute a biological measure in part.[5] Nonetheless, a one-dimensional evolutionary explanation of the differential reproductive rates of the Roman Catholic laity and the Roman Catholic clergy would be quite bizarre. We need not rest our case on such examples, however. There are many reasons for resisting the reduction of social systems to biological systems. The reason that ought to have weight for biologists is that the standard reductions within evolutionary biology itself are unsuccessful (see Chapter 3).

Closure and Method

At this point it is possible to make the search for explanatory closure look hopeless by harping on the potentially endless complexity of any investigated system of systems. But, with a secularized epistemology at our disposal, with a variety of models to exploit in a flexible way in a search for robust results, and with many different closure options to explore, there is no reason why game theory cannot be a useful tool (Levins 1966; Lewontin 1976). It *ceases* to be useful when it is used one-dimensionally in terms of crypto-teleological strategies like those attributed to the rational economic species.

An interesting comparison that nicely illustrates the last point is the following: Bookkeeping entries, personal accounts, or patterns of behavior which apparently indicate that a human agent is behaving as a rational economic man are, in fact, compatible with a wide range of explanations, including the presence of structural constraints of all sorts. No immediate inference is possible from the bookkeeping of rational economic behavior to explanations for the behavior. (This is, of course, to be expected given Garfinkel's theory.) An evolutionary parallel to this point has been made amply and elegantly by Elliot Sober and others (Sober and Lewontin 1982; Gould and Lewontin 1979). Namely, the bookkeeping of population statistics and the apparent adaptational utility of particular traits are not sufficient to allow us to conclude that those particular traits were selected for, in the standard Darwinian sense, though they arose through normal evolutionary processes. Many other potential explanations, consistent with the bookkeeping data, are available. In each case, the temptation to make the immediate inference from bookkeeping data to simplistic explanation in terms of the ideal types "rational economic man" or "Darwinian selection" betrays an *a priori* commitment to the ideal types which is not defensible on any scientific grounds.

Explicitly stated in terms of game theory, this means that a matrix that seems to yield stable optimal strategies in a single game may, in fact, be the consequence of, say, suboptimal satisficing in a complex array of interpenetrating games being played by "players" at several systems levels, sometimes deterministically, sometimes stochastically, all at the same time or during overlapping spans over evolutionary time. The closure condition favoring the explanation in terms of the simple one-level game is usually imposed by brute force, and it explicitly bars us from examining the more complex possibilities.

In a very dialectical way, the most scientific strategies show up the problems of closure best. Experimental design is largely a technique for imposing closure on systems so that clear-cut explanatory connections can be established. When experimental design is best it creates the widest gap between a phenomenon contrived in the laboratory and the phenomenon found in a natural setting. In the laboratory, subsystems are artificially sealed off, and their isolated behavior is studied. It takes a very complex set of inferences to move from what is learned in the laboratory to an understanding of what is going on in nature. The complexity of the move is seldom appreciated. By ignoring the complexity, and by thinking of the natural (and social) world as an enormous laboratory, the results of the narrow neo-Darwinian program can be made to look more impressive than they really are.

A certain sensitivity to the gap between laboratory and nature is shown by John Thompson in the work already cited. Two passages in particular are directly related to my emphasis on explanatory closure: Thompson refers to laboratory experiments investigating inter- and intraspecific competition among snails. He says:

> In the laboratory experiments, interspecific competition and its effect on growth of snails was as intense as intraspecific competition, thereby indicating that interspecific competition could be an important selective force. Also, individu-

als of the same size ingested the same size particles in all Hydrobia species, and individual growth of snails was correlated with the availability of diatoms of particular sizes. Together these results indicate that species with the same size frequency distribution have nearly complete overlap in sizes of food they select and they compete for the resources. What is missing from the analysis, of course, is a demonstration that food is limiting in natural populations. Also the extent to which the populations mix from year to year is unknown. (Thompson 1982, p. 42)

The two concluding sentences are key to my point. The first explicitly states the need for straightforward Malthusian closure if we are to extend the laboratory results to nature. In fact, of course, *every* selectionist explanation based on the competition for scarce resources depends on the demonstration that resources are indeed limitingly scarce. This sort of closure is often attempted by assuming that populations will naturally expand to the point of Malthusian closure. But this is a bizarre assumption. For it in turn assumes that there are no constraints in any other relevant dimension or at any other relevant level which keep the population below Malthusian limits in the dimension being focused on. Now, I have no doubt that laboratory conditions can be controlled in such a way that laboratory populations are under Malthusian closure. There is little doubt that *some* populations in the wild are at Malthusian limits with respect to some given resources. But I seriously doubt that Malthusian closure even exists for any population in all dimensions at any time. To assume so would be to assume that the earth is a thermodynamic plenum, a zero-sum game with respect to all accounting systems. But the earth is a thermodynamically open system, as are all ecological subsystems. It seems probable to me that when Malthusian closure does occur it is usually not as a result of population expansion, but rather as a result of "environmental contraction" in response to climatological, geophysical, and industrial events. The point here is not that selectionist explanations are impossible, but that they must be offered along with joint closure conditions. There must be Malthusian closure in a specific dimension relevant to selection, *and* there must be no other dimension relevant to selection.

We can accept selectionist explanations based on the competition for scarce resources only when Malthusian closure can reasonably be demonstrated. It seldom is.[6] What usually happens is the reverse. We are told that when plausible selectionist explanations can be provided, then we can reasonably assume that Malthusian closure was present. This of course is begging the question unless we have an *a priori* commitment to the exclusiveness and ubiquity of such explanations. The so-called "Just-so Stories" that have become justly familiar in the literature of evolutionary biology (and especially sociobiology) are usually guesses at explanation without the necessary closure conditions having been established. Somewhere along the line it was apparently decided that Mother Nature was a good, frugal, bourgeois hausfrau.

One of the consequences of good—*especially* good—experimental design is that interaction effects are sealed off, as Thompson also points out. Failure to do so results in failure to achieve a cleanly interpretable experiment. This is far from a criticism of experimental methods. It is rather a clear statement of their simultaneous advantages and limitations. Let us quote Thompson again.

> Most of what is interesting about biological communities cannot be pinned, stuffed, pressed onto herbarium sheets, or preserved in alcohol. Knowing the species structure of an assemblage of organisms tells us in and of itself little more than a telephone book tells us about a city. Nor can what is interesting about biological communities be dissected, weighed, separated on starch gels, or centrifuged into supernatant and precipitate fractions. Knowing the internal workings of organisms in isolation from other organisms with which they interact tells us the "how" of life without the "why." What makes biological communities more than lists of taxa complete with details of how they tick are the interactions among the species. (*op. cit.,* p. 124)

Again this is a warning about making hasty inferences from lab to nature. We are confronted with the question of how experimental findings are to be integrated into explanations of natural phenomena. One key to answering the question is to push Thompson's distinction between the "how" of life and the "why." Laboratory experiments tell us how things *do* happen in the lab and how they *could* happen in nature.[7] Any attempt to demodalize experimental findings to apply them to natural situations requires a dialectical return to the conditions imposed on the experiment itself in order to recall the imposed closure which made the experiment possible in the first place. An assessment of these closure conditions is essential to any claim that nature is like the laboratory in exactly the way it must be if the inference from lab to nature is to be acceptable. Thompson's distinction between "how" and "why" is very like the distinction I have been making between "bookkeeping" and "explanation."

Thompson's book as a whole constitutes an argument against inferring from any one experiment to explanation within evolutionary ecology. It is an argument, in short, for the necessity of multiple pathways of access. Nor can experiments simply be added together. Every experiment will impose closure, limit interactions, etc., in a particular way. The assumptions allowing us to say that any graphs derived from two experiments can be combined in a simple way are likely to be very restrictive ones and, of course, have to be justified with respect to their fit with natural circumstances as well as their fit with one another.

Simplisticity

The argument of this chapter is that complexity has to be confronted honestly, and that orthodox neo-Darwinian explanations often fail to do so. This does not minimize our debt to the orthodox Darwinian program, nor does it deny that relatively orthodox Darwinian explanations are bound to have a place in any more comprehensive theory, but it does give selectionist explanations a less central role than they had in the heyday of the synthesis. The orthodox neo-Darwinian reply to this line of criticism could well be to admit the complexity of evolution but to argue that it can be dealt with by taking ordinary natural selection as the standard explanatory model and then treating all additional complexities as modifiers of the standard Darwinian events. Unfortunately, this strategy runs into all the problems of ideal types and *ceteris paribus* clauses we have canvased. In addition, an argument to this effect is usually forced to resort to simplicity as one of the values upon which theory choice is to be based (Kuhn 1962, 1970). In other words, at some stage, anyone who insists on complexity is confronted with Occam's razor.

As it is normally used now, the Occam's razor argument comes down to a scruple. We are to choose between competing theories *(ceteris paribus)* by favoring the simpler one—assuming that the simpler one has an adequate track record as an explainer and (since Lakatos 1970) as a generator of new research. Put this way, the scruple sounds like a sensible heuristic. In fact, however, as many have pointed out, it is problematic in devastating ways.

First we have to remember that "simplicity" is inextricably theory bound, hence that Occam's razor always depends on a criterion *internal* to a particular theory. The simplest explanations for the phenomena *we* like to explain by evolutionary theory are the explanations some of us are faced with in our freshman courses. "God made it (us) that way." When the teacher asks how God did it, he is often told that that is God's business and we ought to stick to what is proper for us to know. Well, there is a long dialectical sequel to this *aporia,* and a short one. The long one involves spelling out the metaphysical and/or existential commitments underlying the two opposing views. No matter how the dialogue unfolds, the invocation of Occam's razor must be absolutely questionbegging.

The short sequel consists of teacher leveling the charge of unscientificness against the student. But this is a dangerous move, for the student must be already inpressed by the wonders of science for the charge to have any persuasive power. Only when the power of the charge is already acknowledged (e.g., when the student's dependence on the fruits of scientifically grounded technology is established) can Occam's razor be invoked. But then the key move is not the razor stroke. It is the entrapment of the student in a commitment to science. This is not an anecdote about freshmen, of course. It is a microcosm of the entire evolutionism/creationism debate.

The competing theories of evolutionary dynamics that figure throughout this book are far less globally antithetical than the faith versus reason alternatives argued by the creationists and the evolutionists. Yet appeals to Occam's razor equally beg the question. One-dimensional selection models seem simple to selectionists, but, dressed up in qualifying caveats until only their mother could recognize them, their simplicity is far from obvious to the unfaithful. *Every* model is a candidate for simplicity—in its own terms. The real issue is which models, handled with which epistemological strategies, are better as explainers and research generators. None of the "sides" in a dispute between theories can insist on the *a priori* authority it would take for Occam's razor to have any polemical force.

CHAPTER 5

Layers, Loops, and Levels

The problem of hierarchy has arisen in several ways over the course of the argument of the last two chapters. It clearly lurked in the wings as we discussed structured emergence of structuring structures. It was explicitly before us in the discussion of possible players in the evolutionary game. Reductionism, units of selection, levels of explanation, theory compatibility or comparability are all somehow likely to generate questions about hierarchy. To my mind, talk of hierarchy in these cases is an artifact of the misspent youth of metaphysics.[1] While thinking in terms of hierarchy has produced some salutory consequences, there are some eventual dangers to consider. After pinning down the structure of the situations that can genuinely (if narrowly) be thought of as hierarchical, I want to contrast other situations—often talked about as hierarchical—with the "paradigm," and suggest ways of talking about them other than as hierarchies. In the end I will take (minor) issue with the work of Stanley Salthe.[2] I admire his work very much, exploit it where I can. On the other hand, I think that Salthe has let ontological zeal clog up the works of his theory and obscure the power of his analysis.

A Genuine Hierarchy

The word "hierarchy" cannot simply be used as a synonym for "strong ordering," as some people tend to do. In other words, I do not want us to say that the numbers 10, 20, and 30 are arranged in a hierarchy from smallest to largest. In fact, we would do well to remember that the word hierarchy is in the same bag with oligarchy, monarchy, geriarchy, matriarchy, and so forth. That is, its primary historical reference has been to systems of authority. I am, of course, well aware that "hierarchy" means *precisely* "strong ordering" in the usage of some fields. I want at least to examine such usage once again.

As a paradigm (or papadigm) of hierarchy let's recall the hierarchy of the Roman church. At the bottom is the laity. Above them is the parish priest, then, roughly, the bishop, the archbishop, the cardinal, and the pope. This is a hierarchy of authority. The authority of those higher in the hierarchy supersedes the authority of those below them with respect to a set of relatively well-defined issues. These issues can be called the domain of the hierarchy. The hierarchy is 'cleanly traceable' *only* with respect to the defined issues, and with the unambiguous specification of the one-dimensional order of the positions in the hierarchy. The totalization of the hierarchy would mean expanding the set of issues to encompass every conceivable issue, and confining the specification of the members of the hierarchy to their designation within it. The hierarchy yields unambiguous judgments of supersession and jurisdiction so long as each member of the hierarchy is uanmbiguously located at one and only one position in the hierarchy.

Like most genuine hierarchies, this one is pyramidal. There are lots of folks occupying the position at the bottom, fewer occupying the next rung, etc., until the top position is occupied by a single person. When questions of supersession are raised *across* the hierarchy, i.e., within a single position, they cannot be answered at that level, but must be appealed to the next level. That is, if the situation is genuinely hierarchical, those at one level cannot act as "judges in their own cases" in matters of supersession. Someone or some institution at the next level up then functions as a source of secondary structuring of the next level down.

Membership in the church hierarchy has a significant effect on behavior. The current (1987) probability of Woytla doing this or that are different than they would be if he were not pope. The same can be said for the probability of everyone in the hierarchy doing this or that.

The degree of totalization of any hierarchy determines the possibility space left to be organized by other structures. Some of these other structures are linked with the hierarchy. For example, there are offices within the College of Cardinals which differentiate roles for cardinals. Or, there are monks as well as priests, and their relations of supersession and jurisdiction are relatively well defined. But other structures are possible which have no linkage with the church hierarchy. For example, suppose a bishop needs to have his hernia repaired. It is likely that he will go to a doctor who is a member of the laity. This doctor is himself, of course, a member of a hierarchical profession, one that is independent of the church hierarchy. And it is likely that for medical purposes the doctor's hierarchy is the one that will structure the relationship between him and his bishop—though forms of address, manner, etc., may be governed by the church hierarchy. The bishop will probably "submit" to the judgment of the doctor with respect to treatment, the necessity for surgery, etc. Expertise tends always to generate both hierarchy and its acceptance by those who assume the bottom positions.

The church hierarchy, then, is a single structure within which members are arranged by the roles assigned them in positions of supersession. This supersession holds within a domain where it is "governing," but it is not totalized. Therefore it leaves space where members of the hierarchy, in their nonhierarchical characters, can act "on their own," or within some other structure.

We could now ask whether all hierarchies are like the church hierarchy—and thereby end up in one of those interminable definitional disputes so beloved of philosophers of modest attainment. I think that the point is not to search for a definition, but rather to examine the similarities and differences between some of those things which have been thought to be hierarchies by someone or other. The outcome we should shoot for is a modest one: not to have a Platonic lock on the form of hierarchies, but to have a clear view of the sameness and difference. I have used the church hierarchy merely as a benchmark to yield the handy provisional criteria of supersession, domain, totalization and residual space. It is worth remarking, however, that the background aura of "governance" of lower levels by higher is never quite absent and accounts in part for the attraction the concept of hierarchy holds for anti-reductionists.

Biological Hierarchy

In applying these thoughts to biology, it is convenient to start with some remarks of Ernst Mayr, who thinks of atom, molecule, cell, organ, tissue, and organism as arranged in a hierarchy. "Complex systems usually have a hierarchical structure, the entities of one level being compounded into entities at the next higher level, as cells into tissues, tissues into organs, and organs into functional systems. To be sure, hierarchical organization is also found in the inanimate world, such as between elementary particles, atoms, molecules, crystals, and so on; but it is in living systems that hierarchical structure is of special significance" (Depew and Weber 1985, p. 57). With the church hierarchy it is the authoritative supersession which makes the structure a hierarchy. Mayr's example, on the other hand, is one of "nested" structures and structures of structures. (And let's not think of the concept of "nesting" as necessarily univocal.) If we were to insist that any hierarchy must exhibit relations of supersession, what could they be, in Mayr's example?

If our basic picture of the situation is a "part/whole" picture—and it is very hard to shake this picture—then we have a strong ordering from little parts of less-little wholes, to less-little wholes as parts of bigger wholes, etc., up to the biggest whole, which is in this case the organism. So far no hierarchy. But it is clear that in the back of Mayr's mind there is a conception of governance in terms of which hierarchy can be constructed or discovered. The mediating word is "law." Despite Mayr's noble fight against the hegemony of physics he still thinks in terms of the discovery of laws being central to science—i.e., he is only semi-secularized. So for him atoms are governed by laws (discovered by the physicist), but the laws are, shall we say, overridden when the atoms form macromolecules. For the rules of combination for macromolecules cannot be derived from the laws "governing" the atoms without additional reference to structure. Similar claims are made for each additional step in the sequence to organisms, or at least we can presume on the basis of Mayr's subsequent discussion that such claims are made.

Thus the "hierarchy" seems to emerge from the sequence of overrides. But there are real difficulties with this. Whatever the rules of combination for macromolecules, and whatever the structural consequences of each of the succeeding larger wholes, the laws governing atoms are surely never overridden. They hold for all atoms no matter what structures they happen to be part of.[3] And this is true even if it is conceded that explanations of macromolecular combination cannot be *reduced* to explanations solely in terms of the laws of atoms. So a hierarchy cannot be constructed in terms of the overriding of laws. Structures do not always have higher authority over their constituents in the way that popes have authority over their hierarchically deployed flock.

A better way to view Mayr's nested structures is as a system of complexly interacting structures. In line with what I have said before, structure exists wherever possibility space is organized, i.e., whenever all "microstates" are not equiprobable, and the new probabilities are systematic. To think of this, again, in terms of changes in phase space, when structure is present, the movement of items in phase space is subject to constraints which change the "degrees of freedom" of the items. Hence once the structuring is identified, the number of independent variables required to describe the movement of the items is changed. It's not as if the items no longer obey the "old laws" that "governed" their behavior before structure emerged or was imposed. Thus, atoms bound into macromolecules behave in ways determined by the same laws that isolated atoms obey. But either they have access to less points in possibility space than the "free" atoms because of the new constraints, or the parameters of the space have changed.[4] From this point of view, if we wanted to start with a "from the atoms up" perspective, then the major job would be to account for why the atoms were bound so tightly and complexly that they constituted an identifiable more or less predictably behaving entity such as an organ, an organism, or an ecosystem. Again we notice that the emergence of structured structuring structures is an enabling as well as a constraining process.

I suggest that we think of a "nest" of structures, each of which imposes constraints on the behavior of the other constituents of the nest as a level-interactive-modular-array (LIMA). In general, no straightforward supersession can be established in a LIMA and the relations between its substructures can be extremely tangled and complex. It is necessary to use coordinated research strategies to gain multiple access to pry out the determinate structures involved and to investigate their interpenetration. Hierarchies would then be sorts of LIMAs. Interestingly, if the church genuinely is a paradigm, then hierarchies are organisations of constituents at a single level ordered with respect to higher level structures. That is, the church hierarchy is a hierarchy of *men* given precedence by the rules, traditions, and practices of an institution, the church. The church is not *part* of the hierarchy.

We can now say that Mayr's sequence of structures constitutes a LIMA. There are rich research techniques for identifying each of the structures he mentions. Each of these structures imposes constraints on each of the others. Clear lines of authoritative supersession cannot be established. Reduction of any of the structures to any of the others is impossible. The concept of a LIMA may not have the dignity of tradition, but it seems to offer a much more congenial frame for evolutionary thinking than the concept of hierarchy. For example, it allows us to think of all the various determinants of evolutionary change together without prejudicing questions of their supersession. We do not have to stop thinking, for example, of "levels of selection" if all we mean to do is point to the variety of structural components of the evolving system. But we do have to delay giving supersession (or independence) to one or another of these structures until the nature of the supersession and the degree of independence are established by empirical investigation. And, if indeed no particular supersession or independence is established, that's all right too. It will loosen the hold of some old dogmas.

Hierarchy and Language

Next we will examine a potential hierarchy which appears in a discussion of Brooks and Wiley's theory by John Collier in his paper "Entropy and Evolution" (Collier 1985). This time we will pay even closer attention to the four considerations which helped us to understand the church hierarchy as a paradigm: (1) supersession, (2) domain, (3) totalization, and (4) residual space.

Collier is concerned with "hierarchical" codes. He says, "An example is the hierarchy of characters, words, and sentences" (p. 14). From a purely physical point of view, these three sets of items are surely related in a straightforward part/whole way. As we have seen before, there is little reason to call this relationship, by itself, a hierarchical one. Collier's subsequent discussion may provide additional reasons, however. For he points out that not all correctly terminated sequences of characters form words, and not all correctly terminated sequences of words form sentences. This obvious truth is important in that it shows us that the *potential* word and sentence producing resources of a language are far greater than the resources actually utilized. "Words are distinguished from non-words by having a meaning or grammatical function, and sentences are distinguished from non-sentences by being grammatical" (ibid.). Leaving aside, for a moment, the point to which Collier is moving, let's see what sort of hierarchy might be present given these latter considerations.

First supersession within a clear-cut domain. What "governs" grammaticality? The rules of grammar presumably do so. Do these rules operate at the "level" of sentences and "govern" the grammatical function of words? Perhaps so. Some respectable theories of language would put the situation in this way. Words, as it were, are not autonomous with respect to the roles they *can* have in sentences. Grammar defines the roles. The fact that most words are limited to a very small number of grammatical roles simply reflects the hegemony of grammar. Only poets contest this hegemony. But characters (e.g. letters) do not fall into the domain of grammar. This means that the domain of grammar cannot be totalized. The degrees of freedom of letters cannot be determined by grammar. The potential of letters to be parts of sentences is not "governed" by the sentences. A good deal of residual space remains potentially to be organized by something other than grammar.

Much of this residual space (though not all) is organized by phonetics. There are vowels and there are consonants, and phonetic considerations determine, to a great extent, the role that letters can play in words. In addition, innumerable events in the evolution of a language provide additional constraints on the ways in which letters can form words. Some of them will be examined in detail in Chapters 6 and 7. From the point of view of *this* collection of constraints, words are partial governors of sentences. For sentences have to contain words, and grammar does not determine which strings of characters are words. So in the domain of wordliness words are hegemonic. That is to say, the level of words has supersession over the level of sentences in the domain of wordliness.

Now, we ought to be careful even at this stage, for with respect to artificial languages, including the one I am using as I type this into my computer, this disjunction between grammaticality and wordliness may not exist. But we now have to face the issue of meaning, and the way it might lead to hierarchy. This is a tangle if there ever was one. For, what a sentence means is clearly a consequence of the words in it; and the meaning a word has is often a consequence of the sentence in which it occurs. There is no clear hierarchy between words and sentences in the domain of meaning. Nor, in general, do individual characters have meaning—though the meaning a word has is clearly a consequence of the word it is, and that is clearly a consequence of the letters it has in it—as the last four words demonstrate.

Again we have to note that "codes" very much like languages can be constructed where hierarchies of meaning-contribution *can* be established—e.g. the propositional calculus as it is usually considered. This is not a very significant fact with respect to natural languages. Furthermore, the positivist practice of taking these simplistic hierarchical codes as the paradigms of language need not be our practice.

Just as it was a reasonable hypothesis that organs were evolutionarily secondary to organisms, so it is a reasonable hypothesis that words are evolutionarily secondary to sentences. But in both cases the *diachronic* hierarchy need not be mirrored in the synchronic analysis in terms of wholes and parts.

So in the domain of meaning no clear hierarchy can be established between characters, words, and sentences. But can the domain of meaning be totalized? That is, can considerations of meaning entirely determine the boundaries between words and non-words, sentences and non-sentences? No. At the very least, grammaticality allows us to talk of meaningless sentences on any reasonable theory of meaning.

To summarize, it is clear enough what the temptation is to talk, as Collier does, of character, word, and sentence as forming a hierarchy. But under the (perhaps overstrict) view adopted in this discussion there is no hierarchical relationship which aligns all three in the same hierarchy. Other hierarchies Collier discusses, especially that of the chemical level, the level of macromolecules, the level of phenotype, and the level of species, could be subjected to the same critique. In fact, they, in effect, have been by Salthe in the course of his own attempts to produce coherently ordered hierarchies.

Hierarchy and Emergence

The basic picture that seems to be emerging is that of succession of constraints applied to an initial system with many degrees of freedom. This, we may remember, is one of the primary motivations for using the notion of possibility space. We can now ask whether this successive emergence results in a hierarchy of constraints. I think the answer is an interesting one, which fits with the notion of emergence presented earlier: the constraints are diachronically hierarchical, but synchronically they do not necessarily "nest" hierarchically, but rather in more complexly interactive ways.

The diachronic hierarchy is fairly obvious from the evolutionary picture. The self-assembly scenario requires that phase-separated systems emerge in a fairly rigid order. Only with the successive availability of newly stable phase-separated modules do the material conditions for the next modulatory phase come into being. Organisms, then, are possible only given the presence of viable cells; multicellular organisms are possible only given the stable presence of single-celled organisms. When, for example, the slime molds do their number and, for all the world, appear to make a transition from a single-celled to a multicelled form, the very highly structured features of the single-celled form, in addition to the presence of the bifurcation conditions, are the material conditions for the transition.

We must be very careful, however, for the hierarchical appearance of structures of increasing complexity, and hence the hierarchical emergence of the constraints they impose, need not mirror some hierarchy that suggests itself from an atemporal point of view. Organs, for example, which we would be inclined to place lower down on the scale than organisms, obviously have the existence of organisms as the material conditions for their evolution. As we look at them now, organisms seem obviously to be at a "higher" level of complexity, for, after all, they are composed of organs. But we hardly think of the evolution of, say, mammals as the result of previously evolved heart, lungs, kidneys, liver, brain, and bones convening inside a mammalian skin.

There have recently been some good suggestions bearing on the heuristics of the dynamics of the succession of levels in evolutionary sequence. In the work of Salthe (1985), and in that of Brooks and Wiley (1986), the important distinction between "initial conditions" and "boundary conditions" is emphasized. (Instead of "initial conditions," Salthe uses the term "initiating conditions." Since this term underlines the active process generating features I too have been stressing, I adopt his term.) Questions about the complexity of the material conditions for the evolution of organisms can now be put in those terms.

Consider a particularly tight mutualism: for example that between an orchid and the insect that pollinates it. On the strict boundary condition model, that is, the usual model of natural selection, the coevolution of the orchid and the insect resulted in the selection of insects whose nectar-gatherine apparatus was most efficient in exploiting the orchid's resources and simultaneously most efficient in pollinating the orchid. Conversely, the orchid flower evolved to efficiently attract the insect and facilitate nectar gathering and pollination. Particular traits, then, of each of the mutualists had as their material conditions the presence of the other in an already functioning ecosystem within which the selection of modifications would take place. For, just as it is a reasonable hypothesis that organisms preceded organs, it is a reasonable hypothesis that ecosystems had to be present as the material conditions for the evolution or particular organisms and/or particular organs.

An exclusive initiating conditions model, on the other hand, would claim that all the material conditions for the evolution of each of the mutualists is contained within the respective genomes. On this view, variations will be pumped out naturally as the consequence of, say, the entropic dynamics of the genome-as-information-system. The presence of a suitable orchid and a suitable insect in the same vicinity is the consequence of happenstance, and their continued association a matter of mutual convenience (one supposes).

The two views have different consequences for those wishing to establish hierarchy as the basic structural category, rather than levels interaction. On the natural selection model the pre-existing ecosystem is unavoidably a material condition for the evolution of special organs in the two mutualists, so it would be earlier in a diachronic hierarchy and higher in a synchronic one. On the initial conditions (genomic determinism) view the hierarchy is established once for all. The genome is the source of all the constraints within which the evolution of the phenotype takes place. So on this view the ecosystem would be hierarchically secondary, both diachronically and synchronically.

One of the surest results of the dialogue between the defenders of the two views has been the demonstration by both that the extreme view of the other is untenable. While each side considers its relevant examples and considerations to refute the opposing view, it seems to me that someone uncommitted to the victory of either extreme view can take much of the debate to constitute the establishment of constraints on a resulting pluralism. It is on the basis of such a pluralistic view—as yet not fully determined, of course[5]—that I make the claim that the establishment of hierarchy is much more difficult than it seems. Salthe, as we shall see, is fully aware of this. It is clear that the genomes of each of the members of an ecosystem impose constraints on the nature and future evolution of the ecosystem. Yet the nature of the ecosystem places constraints on the future configurations of the genomes of its present and future members. A mere glance at epigenetic phenomena similarly suggests that there is a similar interpenetration between constraints at many "levels."

Pathways through the possible, in other words, are determined by a plurality of interpenetrating constraints deriving from many recognizable "levels," looping back and around one another. Of course, as I said before, the levels are recognizable as such because of the multiple of cognitive access we have to the complex phenomena of evolutionary biology. In addition, we cannot lose sight of the elaborate system of partially sealed off phase separations that contribute to our ability to identify levels. We can easily win the right to talk of levels on the basis of differential access and differential stability of the objects accessed. However, we have not yet, at least, won the right to talk of hierarchy—either diachronically or synchronically. The ability to talk of genotype as phenotype, that is the ability to locate the genome in various places within an explanatory structure rather than just one, makes this point immediately.[6]

Hierarchies or LIMAs

The relation of my views to Salthe's is a straightforward consequence of what I have already said. The contrast between my view and Salthe's can be sharpened in terms of the difference between an ontological and heuristic agenda. For the question exercising Salthe, against the background of reductionist projects in the theology of science, is, "Are hierarchies real?" In order to give an affirmative answer to the ontological question, Salthe takes recourse to philosophical arcana from C. S. Peirce through Russell and Justus Buchler to Douglas Hofstadter and Fred Sommers. The result is a tortured discussion with which, I must say, my metaphysical colleagues would have a field day. Along the way there is a very interesting attempt to deal with hierarchies in terms of a Russellian theory of types—interesting because Salthe provides decisive reasons for rejecting the approach. For example, Russell's "axiom of reduction" must be rejected, and without it the theory of types is incoherent.

I would like, as usual, to leave the ontological speculation to the theologians. In my view no injustice to Salthe would result, for *nothing* in his theory as an organizer of research within biology or any other science depends upon the metaphysical speculation, and, further, as a contribution to an expanded evolutionary synthesis his theory is only encumbered by the metaphysical baggage. Along with the ontological theory, and quite frequently in blatant conflict with it, Salthe produces a very good account of the dynamics of multilevel-interactive evolving systems. Like the theory I have presented here it depends heavily on earlier work by Levins and Wimsatt. It is well worth looking at some of the details of Salthe's view to see what is substantively at stake in a decision to talk in terms of hierarchy or in terms of level-interactive modular arrays.

John Jungck has put forward the key metaphor of "lenses" in his consideration of hierarchies or apparent hierarchies. Considering a system in terms of stacked levels is like looking at the system with a succession of lenses of varying magnification. The metaphor leads to an heuristic view whose core is that any specification of a "level" has to be the simultaneous specification of the structures at that level and of the observation and measurement conditions determining your investigative pathway to the level. This fits well with what I have been urging all along. What follows is that any picture of hierarchy is an artifact of limited access. In some cases this will be a matter of the particular experimental design. In other cases, alas, it will be a matter of *a priori* ontological commitments.

So, a primary difficulty in speaking of biolgoical, environmental, or economic levels is the very identification of the levels. This is especially true if the sorting into levels cannot be done robustly with a wide variety of access routes. Salthe is well aware of this, and in his substantive remarks on particular systems (as opposed to his metaphysical gloss) he is very attentive to these issues. For example, he says in a discussion of the problematic conception of "superorganism" (pp. 206–207):

> Today the word has taken on an unfortunate pejorative sense of representing an exaggerated point-by-point comparison with organisms that was not part of it at its inception. Even proponents of hierarchical viewpoints reject its use as having become through misunderstanding a dangerous lightning rod for cheap shots from the reductionist camp. In any case, even without the word the war goes on between reductionists and more holistically inclined ecologists as to the degree of individuality of local ecosystems and communities. The question will be settled, if at all, on heuristic grounds, ...

I would say that precisely the same remark could be made of all the indentificatory decisions involved in discerning levels and entities occupying levels. Both my analysis of emergence and my discussion of the potential players in the evolutionary game argue to this effect. I certainly agree with Salthe that "levels" can be discerned and detached from *particular* heuristic lines as robustness is established. But this is far different from the reification of levels in some finalized ontological scheme. In one of his moods, Salthe seems to be of the same mind, for he accepts, as I do, that heuristically reductionist lines of investigation can be fruitful even despite the failure of a reductionist metaphysics.

If we call a level-differentiated system a hierarchy, we tend to think that the inputs and outputs of all "lower level" systems are integrated so that the result is a set of sufficient parameters at a higher level, and that the same consolidation into sufficient parameters can occur from higher to lower levels. Equivalently, we think of each level as a quasi-autonomous subsystem. One thing we might think about in this regard is that the metaphor of levels is one that we tend to import into a situation *a priori*. In many cases, to think of systems in terms of stacked or nested levels we have to revert to a rigid conceptualization of whole and part we really are not happy with. It may be that my interest in social systems, as opposed to Salthe's interest in biological systems, makes me more sensitive to certain pitfalls. For example, once you have a national currency and a national banking system the stability of individual enterprises cannot be specified completely without talking about the stability of the national economy. But, likewise, the stability of the national economy depends, in part, on the stability of the individual firms. In addition, the stability of individual firms depends on the stability of the partners, shareholders, etc. Is there a hierarchy here with the nation on the top and individuals on the bottom (or the other way around, for that matter)? I don't think so. Firms are parts of nations, but nations are also parts of firms in the sense that national banking, taxation, and so forth must be internalized by the firm in determining the conditions of its operation—so no unequivocal stacking is possible.[7] Now, firms are not parts of nations *in the same way* as nations are parts of firms, so a complex intersection of separate analyses will be required. But no *one* analysis gives robust results, and if the identification of parts and wholes depends on the choice of analytic pathway, then claims to the presence of hierarchy are going to have to be very delicate ones. I concede that the problems here are extremely difficult. But this seems to me simply to set the agenda for the next round of careful scientific study of complex systems. Again, I imagine that Salthe substantially agrees with this assessment.

In a later chapter I will talk about the emergence of systems with respect to the thermodynamic conditions of their origin and perpetuation. Salthe is also concerned to link his analysis with the developing theories of dissipative structures stemming from the work of Prigogine. Dissipative structures emerge from chaotic flux conditions and, under the right conditions, can become tenaciously stable. All biological entities (and, I would say, economic entities) are dissipative structures. The presence of these structures raises obvious questions of level. The most important feature of dissipative structures for our understanding of levels is the existence of stable rate differentials giving rise to (or coming along with) local structure sealed off in important respects from global flux. In other words, we look for phase separation, and when we find it we look for the sustaining and sustained structures (the structured structuring structures). We then try to discern the "mechanisms" involved.

Often these phase separations are what lead us to talk of hierarchy. Following Herbert Simon, as I did in an earlier chapter, Salthe reminds us that the solar system is sealed off—for all intents and purposes—from the high frequency events occurring at the atomic and subatomic levels, so that the Newtonian accounts are adequate to describe its behavior. Similarly, it is (currently) sufficiently sealed off from low frequency interstellar events. In other words, the solar system is phase separated, and it is all but overwhelmingly tempting to reify this phase separation into "levels." We then line up the levels of high, medium, and low frequency processes in a hierarchy. Such a view yields systems descriptions which tease out sufficient parameters yielding fruitful strategies of analysis. Those from the "lower" level are Salthe's "initiating conditions"; those from the "higher" level are his "boundary conditions." We must not lose sight of the structural features of phase separation which make the sufficient parameters robust and stable. Furthermore, most of the systems we are interested in are far less sealed off than the solar system. And when they *are* well sealed off—well phase separated—the "mechanics" of the separation are far more interactive, dynamic, and energy-consuming than the separation of the solar system from the high and low frequency events surrounding it.

Salthe sees the situation very clearly and is very explicit in confining the concept of hierarchy to those systems whihc are substantially sealed off. However, the resulting estimate of how many hierarchies there are is less certain than Salthe sometimes indicates. In fact, he is in something of a running battle with Wimsatt,[8] who is inclined to emphasize the interactive dynamics of phase-separated subsystems. Eventually, as I said, there is no substitute for examining *how* subsystems are interacting; and this must be done on a case by case basis. There is no shortcut. This is one of the main reasons I adopt the more conservative concept of a LIMA, rather than the more committal "hierarchy," as my basic concept.

As a sort of test match between hierarchies and LIMAs, we could think of the following *speculative* (and I do emphasize speculative,) case. While in some contexts the case would be extremely contentious, it is far enough from Salthe's central concerns or mine to serve as a relatively neutral ground for comparing our competing "intuitions" about the eventual fruitfulness of talking in terms of hierarchy. In addition, it points up many of the issues Salthe emphasizes, including our place as observer/participants of and in nature.[9] The "mind-body" problem, as it has long been crystalized in a Newtonian/Cartesian context, urges us to ask whether mental states can be identified with brain states. Leaving aside the questions that are begged in such a formulation, consider the following: The task is to associate something like a belief (call it "the belief that p") with a state of the brain which is, so to speak, the "representation" of the belief that p in the relevant part of an organism. It seems to follow that if two people had the same belief the corresponding brain states ought to be the same too. Furthermore, the strong presumption is that our beliefs, when articulated, affect the beliefs of others.

Now consider the hypothesis that "the belief that p" is a description identifying a macrostate of the believer. This suggests that there is another description identifying an appropriate microstate of the believer (notice how "the believer" now strains against one's Cartesian reflexes), and the two descriptions are to be somehow associated. In just such a way macrostates and microstates are associated with one another in standard Boltzmanian thermodynamics. To pursue the standard analysis, the first question that has to be asked is whether "the belief that p" is a high entropy or low entropy macrostate. In the standard Boltzmanian way this is to ask if "the belief that p" must be associated with one or a few microstates, or if it can be associated with many microstates. The standard assumption in the *philosophical* tradition generating the problem is that "the belief that p" is a very low entropy state. Everything in the literature implies this assumption, and once Russellian logic rather than some other mathematical framework is adopted, the assumption is virtually locked in place. As the saying goes, "mental states" and "brain states" should match up one to one so that substitutional identities can be established.

On the low entropy assumption there seems to be nothing blocking a straightforward eliminative reduction of the mental language to the brain state language, and no compelling reason to think at all in terms of hierarchy. Indeed, under the low entropy assumption, if we insist on barring the reduction we are left with all the classic mysteries of philosophical dualism, including the mystery of *why* the two irreducible states are associated with one another so reliably. Whatever Salthe's project, I cannot believe that this is a congenial outcome for him.

But there is no reason to make the assumption of low entropy association, and currently there seem to be many reasons to reject it. For example, the localization requirement for the assumption seems false on the basis of current knowledge. The persistent ineliminability of intentionality (context-relative identity) seems to press against the assumption. In addition, other fine-tuned macrodeterminate biological states have been turning out to be relatively high entropy states.[10] Therefore, since the purpose of the present exercise is not to "solve the mind-body problem," but to illuminate and question Salthe's commitment to hierarchies, let's assume, contrary to the usual approach, that "the belief that p" is a description of a high entropy macrostate—or at least a lot higher than we might have thought. It immediately follows that the search for a one-to-one mapping of beliefs on brain states is hopeless. Indeed, my belief that p at time t_0, as they say, is probably associated with a different *micro*state than my belief that p at time t_1 is. This holds in spades for attempts to identify two different people's beliefs by their brain state. As compensation, we can at least make sense out of knowing what we believe without knowing what state our brain is in. It is something like knowing the temperature and pressure of a confined gas without knowing anything about which molecules are where, or their individual momenta.

Now, the ability to distinguish micro from macrostates is generally taken, and is taken by Salthe, to indicate the presence of a hierarchy. Furthermore, another important criteria of his is satisfied in that there is no thought that beliefs as macrostates *directly* engage in transactions with brain microstates. Our brains are phase separated from the environment of words as clearly as the insides of cells are phase separated from their environments. There is no smooth holonomic interaction between them. But of course the identification of beliefs as macrostates of which brain states are microstates in no way commits us to an ontological dualism, *even despite the fact that the reduction of macrostates to microstates is impossible.* No one (I hope) is tempted to posit hierarchically arranged ontological realms of temperature and molecular motion on the basis of an ability to identify macro and microstates of a confined gas. The articulation of the various research pathways grounding the two-level analysis require no such ontological gymnastics, and, in fact, as cognitive access to such systems increases so that robust intersections are found, there will be less and less temptation to ontological multiplication. It is one thing to say that there are structured structuring structures that must be attended to in analyzing complex phenomena. It is another thing to say that these structured structuring structures inhabit a different ontological realm than the subsystems they structure do. Quaternarially folded proteins are not after all in a different ontological realm than amino acids. They are just highly structured. Their structure has enormous consequences for their behavior (e.g., their enzymatic activity) and for our possible access to them through assay techniques. Proteins *as structured* become material conditions for processes that bare amino acids by themselves cannot generate. But the ontological consequences are minimal. Perhaps the same is true of minds and brains. Time will tell. But if we insist on the image of hierarchy and invest it with metaphysical significance we are going to be continually dragged back to the classic insoluble conundrums of philosophy. I am afraid Salthe encourages this. Like Salthe, I find the concept of constraints, and talk of constraints imposed by various levels of organization, very attractive, as every chapter of this book will show. But, unlike Salthe, I worry very much about reifying identifiable levels, detaching them from the pathways of cognitive access that make them serious objects of investigation, and baptizing them in terms of ontological hierarchy. The materialism I defined so carefully in chapter 2 allows me room to wait for future advance before making judgments, and the concept of a LIMA gives me a way of talking about modularity, successive modularization, and the consequent patterns of interaction without invoking hierarchy as the dominant image. No such image is necessary to firmly establish the existence and importance of highly structured entities.

I see no objection to talking of macrostates and microstates in contexts where this conceptualization bears fruit. I do see problems in thinking of the distinction as a hierarchical one; and I have lots of doubts about how far the conceptualization can be fruitfully extended. Even when we confront hierarchy in the full sense, i.e., where we do find dominance and supersession, we have many hard questions to ask. This will be especially true when we discuss the "hierarchical" levels that are supposed to be generated by and in dissipative structures. And the screw is turned that much tighter when these dissipative structures are social structures. For example, do increased population density, the need for pseudo-flux to generate wealth and status differentials, etc., push toward the reification of new levels of hierarchy, thereby subordinating the individual *human* level to *social* levels? And thus, for example, diminish human freedom? If the emergence of structure can enable as well as constain, no *a priori* judgment can be made here. But talk of hierarchy seems extremely dangerous.

Toward a Broadening of Focus

As we move into territory where social constraints begin to intermingle with biological ones as structured structuring structures, it will be all the more important to take great care with the concept of hierarchy. First, social systems are sometimes (but by no means always) the locus of hierarchies in the most literal sense, and their presence or absence is a serious matter. And second, one of the reflex dogmatisms at stake in the connection of biology and society is the "successive overlay" picture of the emergence of culture, politics, and morality. This picture is generated by our old friend, enlightenment Liberal political philosophy. On the Hobbesian version of this view, as we have already had occasion to mention, morality and "civil society" are laid down over the bedrock of the natural man as beast. The natural, bestial, life of "biological man," as we would now say, is the default drive as inertial baseline. That is, on this view, "man" in the state of nature plunges straight ahead in the pursuit of individual self-interest until some force or forces act on him to deviate him from his inertial path. These forces arise as scarcity and danger make the natural life very difficult. In response to these forces, groups of people make instrumentally rational decisions to enter into conventional agreements (contracts), creating an overlying structured structuring structure, namely society. Human life is constrained by these overlying structures, but always discernible beneath them is the default drive, constrained but unmodified.

The alternative view, found in the thought of Calvin, Locke, and Rousseau, is that a natural goodness, or capacity for goodness, is first "overlain" with corporeal existence, which, in turn, must receive the overlay of civil society (again as a conventional, contractual agreement) to allow the capacity for natural goodness to flourish.

Whatever the relative merits of these views as legitimating political mythologies, and devices for accommodating the Judeo-Christian conception of "man," they are both lousy science. They both presume, *a priori*, that social structure and biological structure are radically different in kind—one conventional, the other natural—and that the latter is the inertial baseline for the former. They both result in a linear additive hierarchy of structurings that can analytically be peeled back layer by layer, exposing ever more basic features of being human, until a natural core is reached. They do disagree about the core.

A full critique of this picture requires us to examine the basic distinction upon which it rests: the distinction between nature and convention.

CHAPTER 6

Nature and Convention

In this chapter we assemble a series of considerations that help us to deal with the discontinuities—or reputed discontinuities—between the realm of the indisputably biological and the realm of the human. The discussion is organized in terms of the distinction between nature and convention, a distinction that persists as an organizing rubric for various facets of human life. I am quite sure that some perplexities about this rubric will remain, for the history of the distinction is rich, complex, and tangled. What follows is a cut across a vast terrain.

The Tradition of the Distinction

One of the most common ways in which the distinction between nature and convention has been made is in terms of "the laws of nature" and "human law."[1] We have already had occasion to note the pressure upon us in the Western tradition to "discover the laws of nature." Furthermore, the dualism of natural law and human law is historically familiar. We have to remember that for centuries, in our tradition, the bridge between the laws of nature and human law was precisely the concept of natural law, either as the pronouncement of the will of the Judeo-Christian god, or the pronouncement of reason.

Much of this, as it comes down to us (as the material conditions for our own thinking), has its roots in the concept of "cosmos" as it figured in the thought of the ancient Greeks.[2] The overriding question we inherit from them is whether nature is a cosmos, that is, an ordered harmonious knowable whole: what we would call a rational system (cf. Finley 1963, chap. VI). And if nature *is* a cosmos, the job of the thinker is to find the foundation of the unity, harmony, and order. There is little need to argue that this search for the foundations of unity of the cosmos has joined with the theological concerns considered earlier to constitute an agenda-setting imperative of Western science ever since. Extravagant attempts to provide overarching principles of a unified cosmos are likely to be termed Spinozistic (or Hegelian), but Einstein's search for a unified field theory, attempts to unify physics in terms of a generalized dynamics, and the more reductive versions of sociobiology are all motivated by a search for unity.

In the face of the near ubiquity of the search for cosmos, we sometimes forget that it is an open question whether the world is a harmonious whole in any particular sense, and it is bound to remain an open question until we "know everything"—whatever that would mean. The claim of total unity is a claim of omniscience. This is true even when the unity advocated is a unity of "method" as in the unity of science program. For the claim is made that a method (a certain set of canons and techniques) is sufficient to produce all knowledge. We must not lose sight of the typical fate of unifying theories. In retrospect they often turn out to have achieved unification by exorcising various items of our experience which resist unification, hence must be illusory (or otherwise non-cosmic) phenomena. In some cases the recalcitrant phenomena are fairly permanently abandoned; in other cases they return to discredit the synthesis that excluded them. For instance, I doubt that many readers of this book hold to a view of cosmos as it would have to be if astrology were a reliable guide to action, or if demonic possession were something we needed to guard against, yet they were integral parts of earlier cosmologies.

Not all rejected phenomena are as successfully banished as the demons. And other intended victims of unifying efforts have more sticking power than the zodiac. Until the Enlightenment, nature would not have been considered a cosmos if the realm of human affairs, human morality, and human politics had been left outside the unity. Yet mature positivism, in its advocacy of the unity of science, left human affairs to an entirely ambiguous fate. Depending on the strand of positivism, human affairs are left to the realm of emotions, to spontaneous natural outbursts, or to the realm of "convention." The first two views are consistent with a thoroughgoing naturalism; the last gives us one of the main reasons for reexamining the distinction between nature and convention. The first two views withdraw any of the import of morality and politics from the realm of the cosmos. The last proposes an interesting dualism which leaves nature, the subject matter of the natural sciences, on one side, and human institutions, practices, and norms on the other, with an analysis of "convention" as a unifying bridge. This vexing dualism has drawn the attention of positivism since its very early days. Currently sociobiology, Skinnerian behaviorism, and some strands of cognitive science offer programs intended to overcome the dualism. All three are widely held, widely discussed theories, attempting to establish a cosmos in their own terms.

Next we have to recall that the concept of nature has always had a central but unstable place in all discussions of morality in Western society—Greco-Judaeo-Christian society. This means that the attempt to come up with a "morally neutral" concept of nature will be extremely difficult in any case. In Garfinkel's terms, the concept of nature generates one set of contrast spaces if it is thought of as the neutral background to human life, and another if it is thought of as a morally relevent participant in human life, that is, for example, as a source of moral imperatives, or something deserving respect and care. Of course special difficulties arise as thought turns to "human nature," for in this phrase more than any other reside the ambiguities between nature as essence and nature as the wilderness from which we emerged. Are the nature of man and the man of nature one and the same? The Hobbesian thinks so; others dispute him. Still others (including me, I think) cannot make much sense out of talk of essences and find nothing but mythology in the depiction of the natural man. This is one of those difficulties which outruns the best intentions of well-meaning investigators. For reasons discussed in chapter 2, science does not have full control over the way its activities and results are integrated into the social world. Thus, while within the scientific ethos a concept of nature may be adumbrated without reference to any moral or socially relevant considerations, as the scientific subculture is integrated within the rest of the culture the chosen concept of nature takes on moral relevance. The sort of dialectical struggle that ensues is exemplified in the evolution/creation controversy.

As nature and convention are discussed in open societal contexts, one of the major themes that immediately emerges is that of nature as a source of limits, constraints, and enablements. What can we expect from ourselves and our fellow humans? What social orders will we be able to establish? What, in our characters, can we do something about, and what are we stuck with? These questions are the classic ones (we might even say "chronic" ones) and they do not go away in the face of science's neat conceptualization of nature. Instead, they persist, generating complication after complication in any attempt to develop any human ethology or sociobiology. We have seen this amply over the last few years.

Consequently, the territory we need to traverse is not easy territory, given all the cross currents, competing discourses, and ideological freight. Because of the tangle of issues within which the distinction between nature and convention is caught, I think we have to emulate the developed sciences and try to find something like a laboratory experiment. We need to see a simplish, scaled-down system within which the distinction between nature and convention has a place, but which is not too complicated by "side issues" to understand. Fortunately, I think we can find such a laboratory surrogate. It will have to be worked out rather carefully and in some detail, but in the end its contribution to our understanding of the discourse of nature and convention will make the effort worthwhile.

In pursuit of such a laboratory surrogate I offer an analysis of an inconsequential human practice which is dialogic, goal directed, and extremely limited. Bridge, especially as played by experts, is the most intellectually complex of card games, especially from an information-theoretical point of view. (I will present only the bare bones I need to make my points.)

Bridge: The Game

Bridge is played between two partnerships (N-S and E-W) with partners sitting across from one another—a first constraint on information space, as we shall see.

The play of the hand results in the capture of some number of thirteen total "tricks" by one partnership, and the remainder of the thirteen by the other partnership. A trick consists of four cards, one from each of E, N, W, and S, played in turn. Players must play a card of the suit led if they have one. Rules of capture are simple: the cards in each suit are ranked from deuce to ace, and the highest card played in the suit led captures the trick—with one exception. One suit can be designated trump. Then, a trump captures a trick no matter what rank card of the suit led appears in the trick. A trick in which more than one trump is played is captured by the highest trump played.

The play of each hand is preceded by an auction which determines either which suit will be trump or that no suit will be trump. The auction also determines the number of tricks one of the partnerships contracts to capture. The scoring system is elaborate, producing complexly discontinuous desirabilities of contracting for various numbers of tricks. We will not go into the complexities except as they determine the vocabulary and boundary conditions we examine.

The contracting partnership contracts to capture six tricks (the offensive book) plus from one to seven additional tricks. This establishes the size of the vocabulary of the auction. The auction consists of each player in turn, starting with the dealer, making a call chosen from the following vocabulary: "pass" is a call; "double" is a call; "redouble" is a call; the other calls consist of a number from 1 to 7 plus either "clubs," "diamonds," "hearts," "spades," or "no trump." There are thus 38 calls in all.

There are $52!/(13!)^4$ possible deals. Any bidding system can be thought of as a language constructed out of the 38 calls in order to manage the exchange of information between partners seeking optimal contracts in as many of the $52!/(13!)^4$ deals as possible. Optimality is fairly rigidly defined in terms of the scoring system, but a few caveats have to be mentioned which materially affect the concept of "available information space" and subsequent considerations of rationality. First, from a "perfect information" point of view, optimal contracts could be defined in terms of the actual trick-taking capacities of the four hands, assuming perfect play by all three players (declarer plays his hand *and* that of his partner which is exposed; defensive players play only their own hand). But bridge is an imperfect information game in the play as well as in the auction, so optimality is more reasonably defined in terms of statistically weighted *expected* trick-taking capacity. So defined, optimality becomes dependent upon the circumstances of the game. Some risks are worth taking in some circumstances but not in others. So optimality is weighted *value* dependent and weighted *context dependent*. But, finally, in all real world circumstances the assumption of perfect rationality ought to be weighted too. So in many actual cases additional risks may be reasonable, weighted with respect to the skill of the opposing players.

The Information Space

Any bidding language arranges the bidding space with respect to all these considerations. In fact, not surprisingly, the most usual languages are geared to the central range of most probably ("normal") configurations, and the abnormal hands are left largely to the skill and imagination of the players. This too affects considerations of rationality, since in conjunction with the one/many, many/one relationship between auctions and deals, it guarantees not only that the auction and the deal must be connected in, at best, a relation, but that there are indeterminacies at the edges of any such relation. These indeterminacies could probably not be repaired without eroding the efficiency of the language for the normal range of hands.

There are rules about the order in which calls can be made which severely constrain the information space of the auction. The call "pass" may be made by any player at any turn. Leaving aside "double" and "redouble" for a moment, the other calls are rigidly ordered in the sequence "one club," "one diamond," "one heart," "one spade," "one no trump," "two clubs," "two diamonds," etc., up to "seven no trump" as the last call in the sequence. A "sufficient bid," that is, a legal call, must be a call later in the sequence than any preceding call, though it need not be the succeeding call in the sequence. "Double" is a legal call when and only when the last call other than "pass" was a call by a member of the opposing partnership from the sequence "one club" to "seven no trump." "Redouble" is a legal call only when the preceding call other than "pass" is "double" by an opponent. The auction is at an end when three successive players make the call "pass" after some other call has been made. The last call before the three passes that close the auction determines the contract. The partnership which has made the last call from the sequence "one club" to "seven no trump" has undertaken to capture the number of tricks denoted by the call in the suit denoted by the call (or in no trump). Thus if the final call is "four spades," the partnership has contracted to capture ten tricks with spades as trump (offensive book plus four). If the final call is "three no trump," the undertaking is to capture nine tricks with no suit being trump. And so forth. Thus the basic rules of the discourse—the information space—are established.

A partnership is most likely to reach their best contract if the information exchanged in the sequence of calls allows them to determine the optimal trick-taking capacity of their combined card holdings. Thus each call must be as finely tuned as possible to convey information yielding an accurate judgment about the optimal contract. There are many bidding languages, some in use, some obsolete. Since they all use the same 38 calls, you would suppose that they were, in effect, dialectical variants, and so they are. Furthermore, given the narrow uniform common purpose locked in by the rules of the game, you might think that all the languages were related in such a way that questions of mutual translatability would not raise serious problems. But this is not so—for interesting reasons.

There are calls in some bidding languages that do not and could not have an equivalent in other systems. This is a consequence of context dependence. Nonetheless, the two nonequivalent systems can, in general, offer the same informational resources. The information space is arranged differently in the two languages, but the result is the same. One of the sorts of device for producing this situation will lead us back to our main concern.

There is a well-established "metavocabulary" in the bridge world which distinguishes natural from artificial bids (calls) and, among the latter, conventions. These distinctions occur, we ought to notice, within an activity which as a whole is highly conventionalized and artifical—an esoteric game given its present form in the 1920's. Even so, it still makes perfect sense to talk of some bids as natural. To wit, a bid is natural primarily if it expresses the judgment of the bidder that the contract potentially denoted by the bid is a reasonable one for the partnership. So if I, as dealer, were to pick up my hand and open the auction by saying "four spades," I would primarily express the judgment, based on my card holding, that my partnership has a reasonable chance of capturing ten tricks with spades as trump. If my partner in her turn says "six spades," then she is expressing the judgment that if my judgment was reasonable, then given *her* holding the partnership has a shot at capturing *twelve* tricks with spades as trump. Both bids are natural precisely in the sense that they express reasonable judgments based on the information in hand (as it were).

But such are the limitations on information space that in many auctions, especially among the best players, the only natural bid in the narrow sense is the one directly preceding the three passes that close the auction. For example, the call "one club" has the natural meaning "In my judgment, based on the fact that my hand is of significantly above average expected strength and the fact that it contains a club suit of above average length and strength, our partnership has a reasonable expectation of capturing seven tricks with clubs as trump." Once in a while, nowadays, the call "one club" actually does have its natural meaning. More often, however, it has a meaning designed to use information space more efficiently. It has the meaning, "I have a hand strong enough to open the bidding; possibly but not necessarily a strong club suit; almost certainly no strong diamond suit; quite possibly a heart or spade suit (or both) containing four cards; and I want to embark on the search for the optimum contract."

Now, if we look at the evolution of this new meaning of "one club" as a response to the need for efficient utilization of information space, some interesting considerations emerge. First, the change of meaning places new obligations on the partner of the player who has made the call. If I were the partner of the player who had bid "one club," and *if I knew that the bid had its natural meaning,* then I could, in my turn, pass if my own holding confirmed that the partnership was best off contracting to capture seven tricks with clubs as trump. But if within the partnership the call "one club" had the new meaning, then I would be obliged (though not compelled) to continue the conversation in search of the optimal contract. My partner's call of "one club" presupposed a continuing exchange of information; without that presupposition it would be impossible to utilize "one club" artificially but efficiently.

The possibility of a continued information exchange, coupled with the superior efficiency of using certain calls with an artificial rather than a natural meaning, quickly led to the nearly universal adoption of what are called "approach forcing systems": bidding systems which utilize available information space by attaching obligations to succeeding calls by partner. They allow information to be conveyed because they detach calls from their narrowly natural meaning with some assurance that subsequent opportunities to convey information will be available. This assurance depends, in many cases, on the agreement that a certain call by one partner *forces* the other partner to make a call other than "pass." Indeed, in some cases, the forcing bid rigidly specifies which of a very few calls a partner may make, and that he *must* make one of them. Freedom of information exchange thus imposes obligations. Interestingly, sometimes an optimal contract would be reached only if the obligation were disrgarded. However, approach forcing systems presuppose that such cases are infrequent enough to be worth the cost. Bridge players believe that there is a special place in Hell where partners who pass forcing bids are sent.

We have to pause here for a moment to appreciate something that veteran bridge players have already noticed. Once approach-forcing systems have become entrenched, players refer to the meaning of a call within the (generalized) approach-forcing system as the natural meaning. In fact, veteran bridge players have probably felt somewhat uneasy with the account of natural bids I have presented so far, because the information space of approach-forcing systems has become second nature for them, as has referring to natural bids in terms of approach-forcing systems. Here, then, is an example of the working of structured structuring structure so clear as to become nearly a parody. A complex of intelligent decisions about the efficient management of the structured possibilities for information exchange leads to the reorganization of that space by means of imposing constraints. These constraints, almost in Matsuno's words (Matsuno 1984), reduce the degeneracy of the information space; tighten the limits of possible bidding pathways; determine those pathways to routes of maximum information exchange. The effort is so successful that they become part of the nature of bridge. No one learns the game anymore except through the approach-forcing restructuralization. It becomes part of the nature of bridge, and determines future pathways of exploration of the management of the information space of which it itself has become a major structuring structure.

An especially important feature of artificial bids is that they deliberately introduct vagueness or ambiguity in place of the determinate meaning of natural bids. "Vagueness," from the same root as "vagare," refers to wandering about indeterminately, "Ambiguity" refers to indecision between two or perhaps a small number of determinate alternatives. The judgment involved in the introduction of vagueness or ambiguity is that, given the opportunities and constraints of available information space, a bit of vagueness or ambiguity will further the aims of the partnership dialogue more efficiently than the determinacy of natural bidding. Later contributions to the dialogue are expected to clarify what was vague or ambiguous and to justify it as a more useful basis for ultimate success than determinacy would have been. It is important to note here that a major feature of the material conditions which make it possible for vagueness to be more useful than determinacy is that deals and auctions are connected by a relation and not by an isomorphism or a function.

Additional dimensions of meaning can also be utilized to make the bidding dialogue efficient. The Italians dominated the world of international tournament bridge for years. Several of their best partnerships talked of each bid having a *"tempo." Tempo* is hard to explain, impossible to define, and translatable only with exquisite sensitivity, but what was meant was that as the partnership dialogue developed each call was chosen for more than its natural or even artificial meaning. Every call is a choice among several possible calls, each almost right, but not quite; each is justified on the basis of the card holding; each moves the dialogue forward, but one *urges* it forward, another cautiously *allows* it to go forward. And then there are the calls that say "Stop! I think we've gotten ourselves in trouble!" The choice between these calls is a choice of *tempo*.

The scarcity of precious information space in the auction leads to another sort of call whose examination will allow us to draw interesting conclusions later. This sort of call is the preemptive bid. This call is made in a deliberate attempt to "take up bidding space," that is, to deny opportunities for information exchange to opponents who are seeking to approach their optimum contract on the basis of the most accurate information they can provide one another. If I as dealer open the bidding by saying "three diamonds," then I have made every call "below" "three hearts" unavailable to my opponents (and to my partner, of course, but the risk may be worthwhile). The opponents' exchange of information may well have to take place beyond their optimal contract, or, if not, then I may still have denied them the space within which they could *find* that optimal contract. Preemptive bids are generally artificial in the sense already explained. When I say "three diamonds" preemptively I may have serious doubts about the ability of my partnership to capture nine tricks with diamonds as trump. Indeed, I may be quite sure that we *cannot* capture that many tricks. My judgment is that the impending penalty is better than the result of allowing the opponents the information space they need in order to find their optimal contract.

Conventions

We can now move on to conventional bids, or "conventions," as they are called. Bidding conventions are a subset of artificial bids which have gained a precise meaning and which are so useful that they have found a permanent place in some, most, or even all bidding systems. We need look at only one, the "take-out double." The natural meaning of "double" is, "In my judgment the opponents can't capture the number of tricks for which they've just contracted, and I wish to double the penalty for their failure." In compensation, the contracting partnership, should it succeed in capturing the requisite number of tricks, has *its* score doubled. "Double" is an interesting call in that it takes up no space. If an auction starts "one club," "double," then the next call can be "one diamond," just as if the call "double" had not been made. Furthermore, as they say, there is no future in doubling seven trick contracts. The points potentially to be gained by the doubler's partnership do not justify the risk of points to be lost if the opponents fulfill their contract. So "double" is a virtually useless call as a natural bid at low levels. This makes it available for other uses. Very early in the history of contract bridge it took on the meaning "Partner, one of the opponents has bid "one" of a suit. I have a hand at least as good as hers, and I think we ought to find a contract in one of the other suits. If you have four cards or more in one of those other three suits, then bid that suit." A "double" of that sort is usually treated as a forcing call.

The take-out double is a convention. Like all other conventions in bridge it takes a convenient, useless, or almost useless call and puts it to good use. Calls which are very useful when natural are not available for the assignment of conventional artificial meaning. Calls whose natural meaning is useful in a particular sequence of calls cannot be conventional in that sequence, though they can be conventional in other sequences. This leads us to think of the sense in which conventions are arbitrary. The question of arbitrariness arises because in many discussions of conventions, especially in philosophy, arbitrariness seems to figure as one their characteristic features. "Arbitrary convention" sounds very like a redundancy. But bidding conventions are arbitrary only in the sense that the artificial meaning assigned to a call is not an extension of its natural meaning, but, rather, a *replacement* of its natural meaning. In every other sense bidding conventions are far from arbitrary. The choice of "double" as the call to be conventionalized is not an arbitrary choice, but rather is to be argued for on the basis of straightforward features of its role in a highly structured information space. No other available call could do its job efficiently. Its job needs to be done.

Now, it will have occurred to every clever practitioner of the philosophic arts that instead of conventionalizing old calls we could invent a new call and insert it into the bridge lexicon. What a pathetically trivial thought. Of course we could, for the vocabulary available within bridge as we know it is radically lean. But, then, that is the whole point of bridge as opposed, say, to whist. The challenge of the game is to accept the lean vocabulary with its rigid constraints, and to shape and manage it so that it gains the capacity to do its limited job elegantly and precisely. *Of course* we could expand the vocabulary; and along those lines we could go on to make bridge a perfect information game. That is, we could deal the cards face up instead of face down and go on from there. Incidentally, *that* game is called double dummy.

Creative improvisatory strategies for dealing with interesting contingencies (an unusual distribution of cards, etc.) concocted by bridge virtuosi in the heat of the game are anchored in the stabilities of the natural bids. They exploit natural expectations at the same time that they intend to disappoint them. Partner will be made to "sit up and take notice," for the routine pathway through bidding space has been left behind. "Unusual no trump" is now a widespread convention. It started out with the use of a call to which partner *had* to react "Oh, partner, you *can't* have meant what you just said." Again an improvisatory success starts us on the road to restructuring, for conventions are the halfway point to nature in bridge. Approach-forcing systems were once conventional too; now they're the defining locus of natural bids.

And Convention

It is time to move on to some of the extensions of these thoughts into broader contexts. First, let us reexamine an old warhorse. It is a matter of convention which side of the road we drive on. A convention having been established, it is sheer madness to drive on the left in most of the world. Elsewhere, driving on the left is an unnatural act performed by consenting Englishmen in the privacy of their own empire. Are the driving conventions arbitrary? Well, driving down the center has too much natural meaning to be an available choice for conventional meaning. Nor could we decide by convention to drive on the right while traveling north or east but on the left when traveling south or west. The constraints here are not so rigid as to lock us into a single candidate for conventionalization, and neither of the two *genuine* candidates has any significant advantage over the other. The indifference between the two sides of the road, however, is the *only* arbitrary element in the convention we choose to adopt. That particular bifurcation is cost-free. Other bifurcations would require more or less extensive revisions of the practice of driving.

Next we can reflect on married couples of long standing who have adopted conventions to alert one another that the time for leaving the party has arrived. They *could* adopt the convention of shouting "Fire!", or maybe "Mayday, Mayday!" when they wanted to leave. Instead, if they have any social sensitivity, they adopt other conventions. A circumscribed acceptable vocabulary is available to them. They may choose to adopt an essentially meaningless but common conversational gambit as their conventional signal to one another—e.g. they may inquire about tomorrow's weather. Or, if they are a bit more elegant, they may combine their signal with the beginning of the leave-taking ritual. "Oh my, I still savor that wonderful dessert" may be the chosen signal. The point here is that if the convention adopted is really arbitrary, it is really conspicuous, and tact requires that it be discreet. Furthermore, the partner who may want to stay a while longer needs to be able to receive and respond to the sign without stopping the flow of conversation. The signal to depart is chosen out of a circumscribed stock of words and deeds. It must not be arbitrary. It will almost certainly be conventional. Conventions are almost never arbitrary. There just is not the available information space in human life to allow them the luxury of being so.

Questions of understanding each other or understanding other cultures have lately turned on considerations of convention. Is ritual to be understood as conventional movement and talk? Perhaps it is. But, if it is, then to call it conventional is not to dismiss the possibility of understanding it in as rich a way as we understand "nonconventional" activity, deeply embedded in a culture. If the analogy with bridge is not misleading us, we may look forward to assigning natural meaning, artificial meaning, and conventional meaning to what people say and do—and all this without falling into any Platonic muck. Of course, to assign these various sorts of meaning we would have to understand the practices of the people in question, just as understanding the natural, artificial, and conventional meaning of a bridge call requires an understanding of bridge. So the epistemological circumstances of such understanding will depend heavily on how multiple access can be gained to the phenomena we want to understand. This situation is no different from the ones we have examined in early chapters. Sometimes the possibility of understanding another culture is made to look more difficult than it really is.

Many will still insist that the dominant medium of human action is language, and that language consists of conventional signs connected syntactically, tied to the world semantically, and woven through human practice pragmatically. Furthermore, it will be said, the limits on the number of possible human languages are the limits of human imagination—constrained, perhaps, by depth grammar. Thus the view is that there are many languages capable of doing the jobs we want languages to do. The language we adopt is a matter of convention. The decision is an arbitrary one.

Now, serious study of the constraints on possible human language—rather than uninformed speculation—has only begun very recently. No definitive pronouncements are legitimate yet. But if we leave the theoretical never-never land and ask serious questions about the social evolution of language, its conditions and its dynamics, then fairly solid claims ought to be possible. The number of languages available to a society is very limited. The changes possible *within* a language are also rather tightly constrained. In fact, these constraints are tied to the society's ability to expand its conception of the world, which it does through its investigative and creative subcultures and their interpenetration with the major institutions of the society as a whole. Again, no *a priori* judgment of rate or extent of expansion is possible. These will vary from time to time and from place to place. But that is the whole point. To pretend that all people have all languages available at all times and that the choice between them is arbitrary attributes omniscience to the language users. Short of omniscience, the linguistic space available to people is limited, and their pathways into new linguistic resources are constrained. Real languages are no more conventional than real ways of life. The space available for conventions is limited—just as it is in bridge. And the linguistic choices made as language develops are seldom arbitrary choices. That luxury is enjoyed by very few. This line will be followed out in the next chapter—and, indeed, in all the succeeding chapters.

In the context of proposing sociobiological approach to human behavior and cultural phenomena, any *a priori* distinction between nature and convention is bound to be question-begging. In fact, the *possibility* of a sociobiology based on accumulated understanding within biology and a consequent reassessment of boundary determinations *guarantees* that prior distinctions between nature and convention no longer can serve as established premisses. In general, what we are and do by nature and what we are and do as a matter of convention are questions, as we inherit them, in an historical investigative scheme far too gross to be usable any more. As we saw before when considering more centrally biological material, the useful questions to ask concern the constrained pathways of action, and the sources, within complex interactive systems, of those constraints. Our understanding of these constraints, for all the advances we have made, is still pretty rudimentary.

So, in the laboratory simplicity of the game of bridge, we find an activity with levels and levels of enablements and constraints. Somewhere up in the higher levels a distinction between "natural" and "conventional" has arisen and is a possibly useful contrast to the traditional metaphysical distinction marked with those words. We already have an inkling, I think, of the way the contrast will work usefully. But we have to go deeper.

CHAPTER 7

The Evolution of Second Nature

In describing the bridge auction I have used the concept of information space. The concept may, of course, be problematic. But it was useful as a way of conceptualizing the constraints and limitations on the legitimate discourse of the auction as it proceeds around the table; the sources and varying stringency of the constraints; the room for flexibility and innovation; the uses of univocity and ambiguity to exploit the available resources; and the pressure to locate new resources or to redistribute those already available. Finally, we can say, the example exhibits the grounds of possibility of the distinction between what an act (utterance) means, what it *could* mean, and what it *has* to mean in order to find a legitimate place in a highly stylized mutually adopted discourse.

The Structured Structuring of Bridge

After all, the sequence of utterances "one club," "one diamond," "one heart," "one spade" *could* be the beginning of an inventory of Bluebeard's castle. Our careful account of the bridge auction has shown us the conditions under which the sequence is a bidding sequence rather than an inventory. Or, rather, it shows us the radical transformation of context that would be required for the sequence to be such an inventory. Someone at the table would have to signal an interruption of the bidding very clearly in order to create a new space for "one club" etc. Of course it could be done. We are flexible folk with many simultaneous concerns. And of course the plurality of meanings of "one club" is in *some* sense available to us all the time (or jokes would be impossible); but this availability is not automatic or cost free. Since we are capable of having a language (and a world) of multiple meaning, and since we are not disembodied, dehistoricized, Platonic souls who can escape the practicalities of existence, we have to have elaborate (and creative) ways of enforcing closures on our talk to make mutual understanding possible. The bidding system is almost a caricature of such a system of closures.

So we can ask at this point about the precise array of constraints that allows many of the calls of the bridge auction to be called natural without grossly misusing that word. In other words, we can look at the structured structuring structures constituting the material conditions for the bridge auction. The first and most obvious constraint is the game-defining *telos*. It is the *nature* of the game to be played seriously in pursuit of winning. Second, and just as important, is the accumulated set of rules. They specify what *must* occur in the course of the game, and, more generally, what *cannot* occur. But they specify what cannot occur in at least three senses. The first is roughly what the ancient Romans would have called *flatus in vento*, i.e., an utterance may simply have no place in information space. The second is the sense of what is and is not legal. Illegalities obviously can and do happen, but when they occur they invoke constraints which tightly determine the future path of the game. An insufficient bid, a bid out of turn, or a renege *determines* the space for possible sequels. So, with illegalities, the point is not that they (as acts) *cannot* occur, or that they are misfires, but that they-followed-by-such-and-such-succeeding-events cannot occur. Third we have the sense of "cannot" which refers directly to the strategies for achieving the *telos*. There are those things that cannot be done by rational players hoping to beat other rational players in the long run. Of course these things can and do happen. Some of them are mistakes, inadvertencies, lapses, misjudgments, and so on.[1] But they do not happen to rational players. That is how rational players are defined.

Now the schoolmarmish among us will point out that we have produced a map of "can," "may," and "should" as we often use them. Others will prefer the map "impossible," "illegal," "irrational." Fine. But the point is that these maps (and they occur in quite considerable variety) are historical social phenomena. As such they represent part of an evolved response of a society (or many interacting societies) to a range of experiences and constitute a more or less stable ordering of these experiences. Our microcosmic laboratory analogue shows us the ordering that has evolved in response to a tiny limited set of experiences in an artificially constrained context. In this context, the relevant issues of "can" and "may" concern what players *can* mean, *must* mean, and *may* mean when they make a particular call.

Looking at the artificially concocted information space of the bridge auction allows us to see the results of a succession of structuring decisions which lock us tightly into singularities of meaning, leaving us, indeed, some room for improvement in the techniques of information exchange, and even some room for "pure conventions," but all within a tightly determined space. The historical evolution of the process of determination is available to us—most of it took place only a few years ago (Frey and Truscott 1964).

Obvious in the historical background was the evolution of the particular deck of cards with which the game is played. Then a succession of games was played with this deck, one or more of which ultimately evolved into the game of whist. Once whist was "invented" the material conditions for auctioning off the hand with great precision were established. These conditions closed off the range of ways an auction could proceed (the four card "trick" fixes the upper number 13; the "offensive book" fixes the upper number 7, etc.). Once the auction was invented the material conditions for an elaborate scoring system were in place. With "Auction Bridge" as a mediator, and drawing partly on a French variation of the game, Vanderbilt developed a scoring system which has been amazingly stable, defining a very successful game. Once the scoring system was in place the material conditons were present for the exploration and management of the bidding space we saw earlier. Its constraints and enablements are, as we saw, stringent once the search for optimal results in terms of the payoffs assigned in this scoring system was established as a *telos*.

Once Vanderbilt's game proved to be enormously popular and a significant intellectual challenge, it began to be played for money and at a formalized tournament level. A need for a governing body arose. A "governing" body already existed for tournament whist, and a parallel body was set up for contract bridge. This body began to enforce a uniform set of rules (and a uniform etiquette), and, equally as important, constrained the process of rule change and adjudicated borderline disputes about innovative ways of managing the bidding space. The system of constraints and enablements became extremely fine tuned.

The successive stages of the evolution of bridge are transparent to our scrutiny. The layers of structuring and restructuring can be seen to have emerged in a straightforward way, which can seldom be said of social institutions. Even with a well-documented history of an institution at our disposal, we are usually unable to give an exhaustive account of the possibilities of power and potential legitimations accurate enough to allow us to say what possibility space is available in and around the institution. The matter is complicated by the fact that *apparent* utilization of possibility space often proves illusory. We do not always do what we think we are doing. Furthermore, the tractability of a precise delineation of the information space of bridge is an artifact of its radical one-dimensionality. To think that social institutions are equally tractable would be to make exactly the sort of mistake I decried in my discussion of complexity and closure. For example, bridge is one-dimensional in the sense that *within* the game there is only one major contrast space for explaining why the players are doing what they are doing. Their actions, as it were, are to be understood in only one way, and we can assume that the players all have the same conception of what they are doing. In general, however, human activity is polysemic, subject to multiple interpretations even by the human agents themselves. This gives rise to the often discussed phenomenon of intensionality. Once a plurality of understandings of what people are doing has arisen, our task of determining the boundaries of the information space within which they are acting is made very difficult.[2] This will be an important consideration in later chapters. But, in general, the daunting complexity of social institutions is a primary reason why it is tempting to impose an *a priori* metaphysics of nature and convention on them—but no excuse for it.

The Anthropology of Convention

There is undeniably a significant contrast between conventions as they must be understood in bridge and the conventions talked about in the typical contemporary philosophical account.[3] The conventionalism that emerges within traditional philosophic discourse is a conventionalism of perpetual possibility of initiation. That is, there is strong suggestion that the realm of conventional signs, governed by syntactic canons, is potentially subject to regeneration from an entirely different set of conventional signs governed by a parallel set of syntactic canons. There is an obvious parallel between this view and the *tabula rasa* foundationalist empiricism which must hold that any body of knowledge is potentially regenerable *ab initio* from the appropriate foundational perceptions. That is, traditional conventionalism is a dehistoricized, "god's-eye view" theory. We should not be surprised at this, since at every turn we confront such views in philosophical discourse.

The conventionalism that emerged from our examination of bridge is not a conventionalism of the perpetual possibility of initiation. It is an *in mediis rebus* conventionalism. The arbitrariness in choosing signs and attaching meanings is replaced by the necessity of managing and molding a pre-existing information space. The availability of an utterance for conventional meaning is severely constrained by the prestructuring of a rather small information space. Conventional meaning can be contrasted both with natural meaning and with other sorts of artificial meaning because all three occur distinguishably within a practice to whose information space syntax, semantics, and pragmatics are already endogenous. And this is all true despite the fact that from a god's-eye perspective bridge *as a whole* can surely be looked at as an artificial, highly conventionalized practice.

Near the end of the *Philosophical Investigations* (Wittgenstein [1953], section xii, p. 230e), Wittgenstein says:

> If the formation of concepts can be explained by facts of nature, should we not be interested, not in grammar, but rather in that in nature which is the basis of grammar?—Our interest certainly includes the correspondence between concepts and very general facts of nature. (Such facts as mostly do not strike us because of their generality.) But our interest does not fall back upon these possible causes of the formation of concepts; we are not doing natural science; nor yet natural history—since we can also invent fictitious natural history for our purposes.
>
> I am not saying: if such-and-such facts of nature were different people would have different concepts (in the sense of a hypothesis). But: if anyone believes that certain concepts are absolutely the correct ones, and that having different ones would mean not realizing something that we realize—then let him imagine certain very general facts of nature to be different from what we are used to, and the formation of concepts different from the usual ones will become intelligible to him.

In one way this illustrates the difference between my agenda and the philosophical agenda, at least as it is understood by Wittgenstein. I think there are serious problems with the thought experiments he recommends. Wittgenstein's remarks ought to serve as a cautionary tale for the generators of just-so stories, which, when all is said and done, are no different from the thought experiments of the armchair Wittgensteinian. But this passage and other related Wittgensteinian sayings, including the association of meaning and use, at least have the effect of pointing us in the right direction. First, to state that meaning is use certainly seems to be an attempt to reduce meaning to pragmatics, and it has been understood in this way by some post-Wittgensteinians. Alternatively, however, it could be a dialectical overstatement of the point that pragmatics are internal to linguistic practices rather than external to them. Pragmatics, syntactics, and semantics[4] could all be seen as generating endogenous constraints structuring linguistic practice without any of them being reducible to any other. The structured resources that evolve in any language need not be exclusively the consequence of efforts to mold the language to pragmatic purposes. Indeed, in general, any restructuring of a language of given resource will have to be done in all three "dimensions"—at least to the extent that it has to square accounts with all three. Most restructuring of language probably generates resources which far exceed any pragmatic needs present to motivate the restructuring. But, in general, the restructuring is not simply the restructuring of language—as purely linguistic—but restructuring of discursive practices (in Foucault's sense), the complex amalgams of language and action constituting identifiable segments of social life.

Since there are all sorts of discursive practices, we next have to be very explicit about the limitations of the analogy with bridge in our understanding of *in mediis rebus* conventionalism. (I suspect the same could be said of Wittgenstein's own use of language games.) Bridge really is frivolous, ephemeral, and arbitrary. For many people, life is not. This means that imagining arbitrary alternatives to bridge is very easy, and imagining arbitrary alternatives to life is not. Furthermore, bridge is one-dimensional. The syntactic, semantic, and pragmatic constraints on its information space are all relativized to one frivolous end within one ephemeral practice. Human life is (or can be) multidimensional, and life is, in general, too short for us to afford the luxury of separate language for each of our activities and practices. Certainly we do have some specialized vocabularies for specialized practices, but, as Wittgenstein pointed out, they tend to be the relatively orderly suburbs of a rather disorderly inner city, and the geography of the inner city constrains the geography of the suburbs. Consequently it is extremely difficult to disentangle the countless interpenetrating constraints which have led present-day people to the languages they use.[5] Intra and inter-cultural dynamics have restructured linguistic resources for so long that Whiggish anthropology is as much to be feared as Whiggish history. What is certain, though, is that there never was a time in the history of the species when the information resources available allowed the luxury of *ab initio* conventionalism. Meanings have ever and always been subject to internal syntactic, semantic, pragmatic, and, perhaps redundantly, biological constraints.

Bourdieu puts the point of *in mediis rebus* conventionalism this way: "It is when the social world loses its character as a natural phenomenon that the question of the natural or conventional character (*phusei* or *nomo*) of social facts can be raised" (Bourdieu, 1977 p. 169).[6] That is, something has to happen to "the way things are" to convert them to "the way things happen to be" or "the way things have been arranged." The question of nature and convention conjoins a question about the availability of alternatives and a question about the material conditions for the availability. Consequently, decisions on the "boundaries" between nature and convention will rest on prior determinations of possibility space. So the investigative field here is parallel to ones we examined earlier in biology proper, but with crucial differences because of the important difference in the ways in which alternatives are enabled and constrained. Bourdieu's remark speaks directly to the availability of alternatives. His point is that the existence of an alternative requires a legitimating discourse *and* circumstances which begin to delegitimate, undermine, and desconstruct an entrenched way of life. What these deconstructive circumstances will be, "what it will take" to put the ordinary and familiar at arm's length cannot be specified with any useful generality. Crises are generated in all sorts of ways as multidimensional systems, once stable organizers of society and culture, begin to destabilize in one or another dimension, often precipitating catastrophe or cascade effects in the most unexpected places.

In practice, the plurality-generating history of Western societies obscures the force of Bourdieu's claim. The very phenomena of "progress" and active pursuit of change are layered on top of a history of migration and displacement, colonization and recolonization, conquest and reconquest, which has made uniform and stable culture a "backwoods" phenomenon in Western society. Nearly everywhere there are memory traces of ways of life which have been patched onto an original social fabric. Sometimes the patches actually make up the entire fabric we now see, making it extremely difficult to identify the original way of life.

Furthermore, the impingement of a new way of life on an old one can occur at various loci, at various stages of maturation, and with various degrees of compulsion. Some new "influences" simply brush the surface of a social system. Others—European colonization of Africa may be an example—can, by force of arms and missionary zeal, rip apart long-established, hitherto stable ways of life. This will be discussed in the final chapter. There are also rare examples of invaders and conquerors having become, in effect, assimilated members of the conquered society and converts to its culture.[7]

In Western European society over the last few centuries the rate of cultural interchange has been so rapid, and the history (memory) of the variety of old and new ways so complete and vivid, that coexisting alternative ways of life seem almost ubiquitous. In the light of these co-present alternatives, all ways of life seem equally a matter of choice, a choice based, perhaps, on a collective estimate of what we *ought* to do, or what we *ought* to be. This view of social and cultural alternatives reinforces and is reinforced by the long-standing rationalist tradition stemming from Plato. In this tradition *all* alternatives, of *all* sorts, are conceived to be potentially co-present to the rational mind, subject only to the limits of imagination.[8] Consequently, in Bourdieu's terms, the rationalist West conceptualizes social and cultural alternatives as if we were in perpetual crisis (maybe we are), and any judgment that a given alternative is more "natural" is based on an extensive rational comparison, with all alternatives in the same problematic imaginative limbo until reason removes the "best" of them to more solid ground.

The Western intellectual then looks down with indulgent understanding on those poor souls whose way of life is, for them, perfectly natural; not an alternative among others, but the way life is and always has been. Meanwhile the poor soul is deprecatingly canonized as the noble savage, and his life is contemplated with nostalgic yearning. This ambivalence about "natural man" virtually defines "romanticism" in the last three centuries and displays some of the tensions within the Liberal ethos—some of the *Unbehagen* within Liberal *Kultur*. It can be seen lurking within orthodox sociobiology.

Conventional Rationality

Most questions of nature and convention, dialectically, can arise only in conditions which fill the Bourdieuean bill; that is, conditions under which the naturalness of life and the discourse supporting it are put at arm's length. However, dialectically again, rationalism itself works to have itself accepted as a natural practice, second nature, or, in the strongest versions, first nature: the characteristic practice of the human.

The claims of rationalism are my main reason for using the game of bridge as our laboratory surrogate. For bridge is a *game,* and the current darling of rationalism is game theory. Bridge is precisely the sort of context where game theory is most at home. When the theory is used in evolutionary biology, as we saw, we are plagued with problems of crypto-teleology. In bridge we have no such "problems." The teleology is the straightforward, unproblematic teleology of winning. Furthermore, optimization strategies are pursued, and there are no serious problems in defining the accounting system of optimization. So we can assure ourselves that we are close to the one-dimensional situation that game theory is meant to help us understand. It looks like the perfect place to examine the role of rationality.

To do so we go back to an examination of the information space bridge generates. As we know from even a cursory glance at the newspaper's bridge column, the space utilized in the bidding of any hand can be represented in a matrix four columns wide and with a number of rows equal to the number of calls made by the player making the largest number of calls. The matrix must be read row by row, from left to right, reflecting the order in which the players make calls. Read in this way, the entries of the matrix trace a path through information space leading to the final contract.

Now, at first glance it looks as though we ought to be able to evaluate the rationality of the path simply on the basis of whether the final contract reached is indeed the optimal contract. But two sorts of considerations must be kept in mind. First, there is a few/many relation between bidding matrices and the distribution of cards between hands. There are far fewer possible bidding matrices than there are possible card holdings. When bidding matrices so bizarre that they would never occur among even awful bridge players are eliminated from consideration, the gap between the few possible bidding matrices and the many possible deals gets to be very large. Second, there is a many/few relation between bidding matrices and particular distributions of cards between the four players. The same hand can be, and often is, bid differently by different partnerships, even when all partnerships reach the same contract. As in many cases where many/one and one/many relations exist in the same system (for example in evolutionary ecology or in the genetic code) efforts to evaluate any one possible path as rational must be highly qualified. Many possible pathways are adequate for particular hands; and particular pathways are adequate for many possible hands. (Brooks and Wiley [1986] remark that evolution is not the survival of the fittest, but the survival of the adequate.) No management of the limited information space is possible which univocally matches bidding matrices with hands. There are, however, ways of grouping hands and matrices to provide criteria for *reasonable* matchings. Were this not so, bidding skillfully and choosing a bidding system would be impossible. Determinism is absent here, so skill finds an essential role (cf. Dreyfus 1983).

Exacerbating the problem of evaluating the rationality of a bidding matrix is the fact that the meaning of any call is radically path dependent (second theorem redundant, in Shannon's sense). The meaning of any call is a partial consequence of the path that precedes it. The meaning of a first call is a partial consequence of its being a first call. To pick an example used earlier, the call "double" has its natural or its conventional meaning depending on the path leading to the call.

These considerations lead us to decide that the entire matrix is the minimal bearer of meaning sufficient to be evaluated as rational or irrational. This is so at the same time that individual calls can be thought to be the minimal bearers of meaning. Individual calls may be evaluated as reasonable or unreasonable; and, in fact, we can so evaluate them without seeing all the way to the end of the bidding sequence. We can do so when we understand the syntactic, semantic, and pragmatic constraints on the information space into which the individual calls are projected. Of course, among the constraints are such imprecise factors as *tempo* and the understanding between partners, which make it impossible to formalize bidding systems completely such that the formalization will give an account of any actual bidding matrix. The formalization will at best establish criteria for whether a particular call is within the range of adequacy as this *range* is delineated by the bidding system and the (possibly idiosyncratic) understanding of the system by the partners. Judgments of reasonableness are available to us, but not judgments of unique rationality.

Finally, the judgments of reasonableness or unreasonableness of calls and sequences of calls are internal to the practice of bridge once the closure condition "the players want to win" is in place. What is left to us to decide on external grounds is very little and very global—i.e., whether it is reasonable or unreasonable, rational or irrational to be playing bridge at all.

Thus all criticism of the dialogue consisting of two partners contributing to the dialogue in turn in order to achieve a specific result must be criticism in terms of reasonableness, not unique rationality. The conditions under which judgments of unique rationality would be possible are very severe and are never met in bridge because of the internal structure of the information space which has evolved and crystalized in the game we know and play. In bridge, many reasonable dialogues succeed, and many fail. The only way we could guarantee reasonable dialogues from failing would be to meet the conditions for making them rational dialogues—i.e., establishing the match between matrices and card holdings as a function rather than as a relation. And that we cannot do.

From outside the partnership dialogue, criticism is hermeneutic. It takes the dialogue as text and seeks to understand it through successive attempts to apply critical canons. From inside the dialogue, criticism is dialectical. It uses constantly generated re-understanding of the meanings in the dialogue to modify the projection of new meaning in the continuation of the dialogue. Dialectical criticism allows the participants to change the conditions of the possibility of meaning from within the ongoing dialogue itself.

The preceding discussion means, from a game theoretical point of view, only that *no optimal strategy is determinable as the "best" path through bidding space.* This is not a shocking situation within game theory. It happens all the time. It does mean, however, that *the nature of the game, a particular deal, and teleological closure are not enough to determine a strategy uniquely*—even if the condition is added that the strategy be an optimal one. So in this case, at least, the claims of reason are roomy ones, sketching out a range of reasonable strategies without locking into a unique one or defining an optimal set. Consequently, the nature of rational constraints is very different in this case than it is in the case of deductive logic or in the fantasy of deductive nomological explanation.

In addition, bridge bidding shows many signs of behaving not as a search for an optimal strategy (except, perhaps, under some sophisticated tournament conditions at the highest level), but rather a search for satisficing strategies. For a given hand there many be two or more contracts whose payoffs fall above a critical threshold (two or more plays for "game," for example), and achieving a result above that threshold may be what the players are after, to the point where a search for the optimum just is not worth the trouble. This weakens still further the relation between strategy and outcome.

In the light of these considerations we have to look again at the way in which the demand for rationality generates constraints—the restructuring of structuring structures. I have already said that one of the more optimistic aspirations of rationalists is to make rationality second nature, move rationality of action down into the realm of rigid natural constraints. In order for this to be done, it must be shown that agents are under stringent pressure in a radical win/lose situation where winning depends on locating a rational strategy. That is, rationality is deterministic only under a species of Malthusian closure. This will be crucial when we examine conventions in a more orthodox Darwinian context.

Coming to an Agreement

As a preparation for further examination of the evolutionary conditions of "conventions" and "information spaces," we have to address the question of what makes conventions possible from a slightly different point of view. Conventions are on any account, including mine, the result of people agreeing about something. In fact, conventions are themselves a sort of agreement. So we can ask what constitutes the agreeing that establishes the agreement. That is, of all the things we do in one another's presence, which of them are the agreement establishing things, and how have they gotten that status?

This threatens to start a vicious infinite regress of a familiar sort; one that has been discussed in the context of contracts and the establishment and following of rules. For, if those things that constitute agreeing have gotten that status because we have agreed to make them acts of agreement, then we have a prior agreement on our hands, and it must be accounted for in turn.

Here we have to recall the distinction between synchronic and diachronic accounts from an earlier chapter. Much of the discussion of the potential regress of agreement and the parallel regresses has been an attempt to justify agreements. Instead of trying to determine how agreements are possible, they have tried to account for why they are binding, where bindingness is thought of as a moral category. Here I am trying to get an account of how agreements are possible, which means that I want to elucidate the structure of a diachronic account, not try to derive agreements quasi-synchronically from some set of fundamental moral premises. Many discussions are ambiguous between these two agenda, but I hope mine will not be. A parallel ambiguity infects the phrase "highly evolved" if care is not taken to separate the meaning of the phrase as "having a late place on an evolutionary trajectory," or "recent," from the meaning "better."

Further, the actual explanatory account of the rise of conventions and other agreements cannot be a fable. For instance, if I were to give an account of the rise of the Stayman convention in bridge, I would be obliged to get the historical facts right, which, in that particular case, would not be difficult. In contrast, the origins of, say, human languages are irretrievably lost, and the actual history will never be known for sure. Nonetheless, this does not give us the right to fantasize irresponsibly (nor, by the way, does it give us the right to cop out into a handy Platonism). What we know at the outset is little but important: namely, that the correct diachronic account will not be an infinite regress. In fact, the theory of the Big Bang places a solid upper bound on the length of any diachronic account. Other evolutionary events surely provide even narrower bounds. So while we may never get precise knowledge of the early history of language use, we are not at a complete loss. We know to some extent what that history *could not* be like.

These days the proposed boundaries of language vary from theorist to theorist—partially under the pressure of issues such as the one we are dealing with here. Body language, body hexus, gesture, digitalized electronic output, and other phenomena are all considered to be "language" on the grounds of productive analogies. But if we confine language to that which is wordly and sentency, then we know that the material conditions for the rise of language cannot themselves be linguistic. The search for initial "agreements" has to go back further than the dawn of speaking and writing.

If I were writing a book about the origins of language, I think I would rest *my* account of the structured structuring structures constituting the objective possibility of language on *Bourdieu's* account. Beyond that, let me offer the following remarks about the possibility of agreement.

Due to the increased centrality of language and sophisticated structuring in developed human societies, we focus too soon on our high-order abilities and lose sight of the very basic abilities upon which they rest. So it is hard to shake the picture of agreements as "talked out," linguistically conceptualized phenomena. But behind these agreements requiring the full apparatus of a highly articulated cultural space are mutual understandings whose objective conditions are the cooperative commitment to a shared task. An image that immediately comes to mind is the nature documentary footage of the hunting behavior of a group of lionesses. *However* we may eventually want to characterize the mutual understanding upon which the hunt takes place, that is, whatever we will want to say about the "mental life" of the lioness, cooperative hunting with differentiated strategic roles takes place among lionesses.[9]

Similarly, without committing ourselves to an account of the mental life of a Teamster, we can profit from watching furniture movers. Moving a table through a door requires a certain amount of very mutually sensitive, cooperatively coordinated manipulation of the table. When two philosophers do the job, one on each end (as I have done as an experiment), you find that a lot of talking goes on. In fact they often end up reasoning themselves to the *a priori* judgment that the task is impossible. But two professional movers simply take the table through the door. Experience helps, but the sensitivity to space, weight, and handiness they exhibit really doesn't require long training—especially not long linguistically mediated training. When you are on one end of the table and have to rotate it either clockwise or counterclockwise, you can come to a mutual understanding by means of a gently applied twist, felt by the other person, and responded to by the appropriate rotation. Or, the other person, perhaps more aware than you are of the corners and angles to be negotiated, may suggest the opposite rotation by resisting your twist and gently twisting the other way. If this gets you to reconsider—look back through the door again, etc.—you can get the rotation done without a hitch; and especially without an agonistic disagreement.

But isn't this to say that the "agreement" the movers come to is the "solution to a coordination problem," in just the way a philosopher such as Lewis would want it to be? The answer is bifurcated. Yes, the rotation of the table is the solution to a coordination problem, but not in the way Lewis would want it to be. For it can only be the solution to a coordination problem if the situation can be conceptualized as a "problem"; and this requires a prior structured structuring. If the prior structuring is the result of previous agreement, we are right back where we started from with the regress. Furthermore, the solution to the coordination problem in this case is far from conventional. The space through which the table must pass imposes constraints obliterating the space within which conventions could be established.

If you are looking for a conventionalized practice, try dueling. Here is a highly coordinated, ritualized activity that depends on a background fluency in the use of highly conventionalized signs. Certainly, ask what coordination problem is solved both by particular duels, and by the practice of dueling. But ask also about the objective conditions (involving honor, standing, dignity, etc.) making possible both the problem and its coordination—in such a way, mind you, that the death of one of the parties is a satisfactory solution to the coordination problem. Ask about the structured structuring structures that gave rise to the signs (cartel thrown in air, glove dropped, etc.) that became "conventional." In other words, win the right to talk of convention by actually looking at what you are calling a convention.

Of course, moving a table can be glossed linguistically, but prioritizing the linguistic formulation is like claiming that the color commentators on a football game are primary, and the players secondary. Mutual understanding of shared tasks is available prior to any language in the wordly sense. Language is an enormously successful facilitator; but it is not what life is all about—unless you are a philosopher.

The feeling that language is primary to mutual understanding is an inevitable artifact of the view that the world is a derivation from first principles, a view we looked at earlier. If we hold a view of cultural transmission in which human cultures are first invented (in a linguistic formulation) by intellectuals, then applied, then of course we are going to think of any mutual cooperative activity as linguistically mediated. Similarly, if we think of the realm of rationality as primarily an intellectual realm rather than a basically practical one, we are going to think of our intelligent activity as linguistically mediated. Of course by now a lot of it is. But even now it sometimes is not; need not be; and at the the root of things could not *all* be.

So the roots of conventional agreement are not conventional. In a life space where all our mutual understanding could be conventional, there would be no real reason for forming the conventions, no shared task to ground mutual understanding. But on the other hand, we find ourselves in a complexly structured culture, highly linguistic in its interactions, highly conscious of the multiplicity of *possible* ways of organizing life. Under these circumstances, as I argued a few sections back, the illusion that the choice of ways of life is conventional, a product of intellection, a product of voluntary and conscious agreements, is a very powerful one. And it may be more than just an illusion if we really do have a whole lot of unconstrained, undetermined space within which to roam. It could, after all, end up that it did not matter who we are or what we do.

We can note, to end this section, that if arriving at mutual understanding were always a matter of conventional agreement, then the continuity between the biological and social I am after here would be impossible. The dualism between nature and convention is simply a species of the Cartesian dualism of mind and body, and the corresponding sealing off of the two dualistic realms from one another. This version of the dualism is often couched in the distinction between the realm of necessity and the realm of freedom. I hope that what I have said here makes such a timeless distinction impossible, while, naturally, allowing us to make the distinction historically on the basis of a careful assessment of the possibility space within which we live out our lives.

Conventional Evolution

You would think that since the radical distinction between nature and convention opens up a dualistic gulf it would be particularly unattractive to sociobiologists, who are, after all, committed to closing the gulf. But not so. The distinction is a very convenient one for them, and to see why is to discern the shape of their program.

Sociobiology has either two or three parts, depending on how you count. First, it wants to explain social phenomena in terms of biological facts and conditions. Second, it wants to claim that biological facts and conditions are deterministic with respect to social phenomena (some may think this is a part of the first, hence the counting problem). And third, it is committed to devising selectionist explanations, which must be employed in carrying out the program. It is this last condition that makes the distinction between nature and convention so attractive.

Both Dawkins, and Lumsden and Wilson have attempted to identify cultural entities whose evolution could then be assessed. Dawkins proposes that they be called memes; Lumsden and Wilson that they be called culturgens. There are differences between the two of considerable interest from *within* the sociobiology program itself. From a (rather short) critical distance, these differences can be put aside for the moment. Then we see that the attempt to identify cultural entities is constrained by the necessity of finding objects which fulfill three conditions. They evolve; their evolution is possible because of a suitable variation among them; and they evolve as a response to selection. In the canonical Darwinian way, their evolution in response to selection must be decoupled from their genesis. If it is not, then selectionist explanations with respect to them will never achieve closure. The decoupling, here as in the more strictly biological cases, must be done carefully (Brandon 1985). What a particular cultural entity is, is surely a "causal" consequence of its genesis, just as every phenotype is causally related to its genotype. But, the theory would maintain, just as nature selects phenotypes independent of their genotypes, culture (or nature through culture) selects entities independent of their origin. This is a familiar pillar of Liberal wisdom, Millean in its straightforwardness. It is familiar in positivist philosophy of science in the separation of the context of discovery and the context of justification. "Theories," say, are generated . . . somehow, and enter the arena of rational debate and criticism where they are tested—and where the survivors live on. If this detachment from origins could not be accomplished, then the process of triage at the rational critical level could not be autonomous, and a bastion of the Liberal theory of cultural transmission would fall.

But the distinction between nature and convention, better than any other dualism that might accomplish the decoupling, severs the (say) theory from its origins, thus making nature available to independently play its role as that against which the theory is tested and that by which theories are selected. A continuity between nature and the theorizing of nature by humans, consistently pursued, would destroy the boundaries required for the evolution of theories to look like a Darwinian evolution. The question, put in terms I have associated with Brooks and Wiley, is this: How can nature be the source of both initial and boundary conditions kept separate from one another? The answer is: By opening up an appropriate dualism, e.g., that between nature and convention. For all their reductive bent, the sociobiologists need the dualism in order to retain their Darwinianism.

The distinction between nature and convention, inherited from so many strands of the Western tradition, and deployed to do so many jobs, is just one of the dualisms that always arise at the boundary of the us and the not-us. The distinction simply tries to manage the two contrary intuitions underlying our culture. We are part of nauture, and we are separate from it. In the grip of a Newtonian conceptualization of nature, and the conception of science that goes with it, to retain a sense of human dignity requires such a dualism—as Hegel saw clearly. There must be a realm of freedom and a realm of necessity.

But if we are dissatisfied with the dualism, and unwilling to reduce ourselves to the mechanisms of the Newtonian picture, we have to question the Newtonian roots both in terms of the ontology and in terms of the conception of scientific explanation that has grown around the ontology. That is what I have been doing all along. But carping is one thing and building is another; in the end only the latter is worth doing. I have tried to do some building as I have gone along. I want to do some more building—or at least drawing of plans—with respect to our understanding of our socioeconomic world.

CHAPTER 8

Negotiating Social Space

The discussion of nature and convention has given us a conception of the structural resources available for various activities at any given stage of biological and social development. It has also given us some clues about how these resources operate and how they may be managed. Structured structuring structures at the same time constrain and enable: limit and create future possibilities. Any system—social as well as biological—must in carrying out its history do so within the possibility space determined by the particular history of its structured structuring. This is not to say that restructuring, and even the dismantling of old structures, will not take place. On the contrary, the history of human systems shows constant restructuring. But change cannot take place willy nilly. The opportunities for restructuring, and the means by which restructuring will succeed, are themselves determined by the possibility space.

The Dialectic of Possibility

We were led to "dismantle" the old gross distinction between nature and convention as far too insensitive and ahistorical to serve as a boundary between, say, biology and society. This sharp distinction, oddly enough, attempts to decouple the evolutionary dynamics of social phenomena from their history in order to render them fit for one-dimensional "boundary condition" explanations. The old distinction, in a word, must trade on inherited but outmoded dualisms. We now know, however, that conventions are a luxury made possible by the abundance of space within which new structures can grow. More usually new structured structuring structures must utilize quite narrowly determined space. Ways of thinking, boundaries of the thinkable, etc., have a way of becoming second nature, strongly constraining available paths.

Futhermore, at this stage we have a rather good idea of the sources and stringencies of some of the determinants of social possibility space. Biological evolution made us beings with certain interactive, coordinative, and anticipatory capacities, and not others. Organisms, ecosystems, and perhaps even social systems have an evolutionary path of increased complexification for the more efficient use of energy resources, thus implying the presence of constraints and enablements directly related to the thermodynamics of life. Information canalization in development is one of the fundamental constraints on biological evolution and has social analogues.

Then (in some senses at the other end of a "continuum") we can recognize the ways in which particular esoteric human practices such as games are constituted by an imaginative set of constraints of an information space. In the case we examined, the possibility space provided by a deck of cards and our desire to play intellectually challenging games is determined by a succession of choices on our part. Yet these choices were by no means arbitrary. This suggests one set of ways in which our attempts to coordinate the various aspects of our lives take place within a possibility space made available by prior historical activities. In examining our own activity in this light we are, or course, seeking to identify the material conditions of human artifice, the determinants of the scope and limits of human practice. From this point of view it is very clear how an enormous range of alternatives is consistent with interpenetrating sets of rather stringent constraints.

As an example of a system of practices built within the enabling and constraining space of very specific structured structuring structures, the sciences themselves show us the way in which an information space is traversed, determined, redetermined, and utilized. The canons, norms, mores, and *teloi* operative within the scientific community at any time draw the boundaries of the information space available to practitioners—including the space available for changing the space. Available techniques and materials determine the range of investigations which can be carried out.[1] The very process of which this book is a part is probably as good an example of these dynamics as any. The reformulations of evolutionary theory at the present time must take place within the space determined by the neo-Darwinian synthesis. Modifications must be suggested as modifications *of* that synthesis. Fundamental restructurings must be tested against the underpinnings of the synthesis. An explicit claim to have produced a genuine *alternative* to the synthesis depends for its legitimacy on bettering the solid results the synthesis can claim, and casting doubt on its failures. It must be shown that the alternative can exploit the new positive findings of working biologists more satisfactorily than the old synthesis. It cannot be forgotten that Darwin established the material conditions for a century of biological advance. Over the last half century the neo-Darwinian synthesis has managed that advance: an advance which itself leads us to reject canonical versions of the synthesis as inadequate to support new advances.

A failure to recognize the old synthesis as the material basis of the new would violate all the canons of responsibility which have grown up within the practice of modern biology, and, indeed, the practice of modern natural science in general. My explication of the process of scientific advance in terms of structured structuring structures, material conditions, information spaces, etc., thus affirms those habits of investigation and criteria of success that have grown up within the investigative practice itself. The language may seem strange, but my claim is that the language is necessary for a full appreciation of the power of biology to legitimate and continue its advances. The alternative account, extended derivation, is sustained partly by a residual commitment to a theological canon: a god's-eye view against which to measure scientific success. There is no reason to think that such a point of view can be found.

The Structure of Gender

Now we need a concrete case with social dimensions to work with as we take the next steps to an integrated dynamic theory. Bridge had its usefulness as a laboratory surrogate, but it is too specialized and truncated to offer a realistic sense of normal social systems. So, to get down to cases, we will examine a concrete phenomenon involving the division of labor. We want to explore the conditions for social differentiation and specialization, and the rubric of the division of labor is an obvious one to adopt. It has, in one way or another, been at the center of most of the social sciences over the last century. Recently it has found a strong echo within ecology as researchers attempt to account for the dynamics of ecological change. Naturally, as research is extended through the various divisions of labor central to human existence, researchers will want to be sure that they do not miss opportunities to relate these divisions of labor to the systems of constraint and enablement serving as the material conditions for their possibility. The "how possible" questions are especially important here, for they lay out the foundation for the more deterministic sounding questions (Why *this* rather than *that?*) to which we get pushed all too prematurely.

I can well imagine circumstances within which the sexual division of labor was tightly constrained by the biology of the reproductive and nurturing process; but for some classes of people in some parts of the world these circumstances have become obsolete, and the constraints have been relaxed. We shall see this in spelling out the determinants of the possibility space of the introduction of women into the academic professions. We need accounts of how the relaxation of reproductive and nurturing requirements, traditional constraints on women, was *itself* possible. Interestingly, at least one instance of the structured restructurings generating such a possibility occurred not within the subsystem of relationships between men and women, but within the development of the economic subsystem in a period when it was almost totally under men's control.

Human children are dependent for a comparatively long time. This is yet another biological constraint, one of the sought for "human universals." But we have to see the possibilities of managing or even evading[2] this constraint in order to assess its importance. Someone has to take care of the kids while they grow up. Someone has to be in charge of the moral orthopedics that convert children to adult participants in a complex social world. This is a true in "simple" societies as it is in modern ones, but it takes on a particular form within modern capitalism. Modern capitalism has expanded the realm of economic activity far beyond any previous society's. The amount of time spent producing is far greater in our society than in any other. In fact, in England in the early nineteenth century, a point was reached when nearly every waking hour or nearly every person not among the aristocracy was spent in activities of production. We can leave the account of the rise of such a situation to others and take it as the given baseline against which nonproductive time (leisure) was recaptured. Given the time required in production, the acculturation of children was a low priority. This was especially true of working class children, who were harnessed to the production process at a very early age. But this situation did not last very long. Neglecting the education of children conflicted with the liberal ethos, especially the strand of Christianity requiring concern for their "souls"—that is, their overall moral development on the Christian-liberal model. We have here the interaction of two contending structuring structures.

The ideological structure was strong enough at that time to become an agency of restructuring. So a division of labor which placed young children in the work force at an early age was an easy target for critics of the modern industrial process. The first reorganization to relieve this tension was to assign the lion's share of the upbringing of children to women. (They were already carrying the lioness's share.) Women, after all, were easily identifiable as the primary source of nurture; and there was a long-standing tradition for the conceptualization of women as the primary denizens of the home while men were the primary participants in the public world—which now became the quasi-public world of the production process.

Once childhood was reaffirmed as a period of life when people ought to be free to pursue the long process of education and acculturation, and at the same time the production process was affirmed as the primary human activity, the historical elements of the Western ethos that differentiated men and women in terms of "natural" spheres had to be invoked to establish a division of labor capable of getting both jobs done. It may also have been true, though this is far more doubtful, that assigning women the nurturing and support in the home, and men the directly productive activities in the world, was more efficient. In any case, a division of labor was necessary and was established. It was neither derivable as a theorem from some combination of overarching sciences, as a matter of "nature," nor was it a matter of "convention." Given the available space and the concatenation of constraints, the sexual division of labor was a reasonable response.

A sexual division of labor of some sort or other is one of the more common phenomena of human societies. Such systems in earlier times, and contemporary ones outside industrialized Western society, may or may not be or have been reasonable responses to perceived necessities. The point is that no inferences can be drawn from the presence of these divisions of labor about the causal primacy of underlying biological "determinants" unless it can be shown that the underlying possibility spaces were identically determined. For example, it would have to be shown that the differences between an agricultural and an industrial society were explanatorily irrelevant to the division of labor that arose within them. The possibility cannot be ruled out, but neither can it be assumed. Biologists will recognize that this is an exact parallel to the distinction between analogy and homology of limbs and organs, and that similar research questions are generated in both cases.

Women's Studies

The preceding discussion actually presents an overview of some of the main structured structuring structures whose effects we can examine in a concrete instance. In discussing the sexual division of labor these days we have the advantage of a large mass of good historical material recently generated by people in what is bureaucratically constituted as "Women's Studies." But rather than canvas that material I want to use the very existence of "Women's Studies" as an object lesson in understanding the grounds and sources of a particular division of labor. In doing so I think I can generate an account which exhibits nearly all of the complexity discussed so far. There are biological roots to the division of labor. In assessing the role of these roots, we will be able to see how they articulate with other structural enablements and constraints to produce a dynamic involving the challenge to and defense of an extremely complex possibility space of a local division of labor. The example is an explicitly American one, not for reasons of chauvinism, but to provide the specificity necessary to approach explanatory closure.

The proximate structured structuring structure within which Women's Studies was constituted is the social institution we fondly call "academia." It is composed of subinstitutions such as universities, colleges, museums, research institutes, journals, and professional societies, and is divided still further into disciplines, departments, fields, and divisions. Some of this subinstitutionalization is mirrored in the organization of social sources of funding and legitimation, e.g., NSF, NEH, NIH, and so forth, or, in other contexts, CNRS, CSIRO, and MRC. Through these funding sources, professions, journals, and the publishing industry, academia has become truly multi-national.

This vast academic institution has responsibility for the path of the growth of knowledge, subject to constraints imposed by other sectors of society. We can discern the exogenous pressures to which it responds; we can discern the endogenously determined pathways it sanctions and makes available for its practitioners. We see how it trains its recruits and controls entry into its practice. As academics ourselves we are frequently (all too frequently) pressed into service to formulate and reformulate the rules determining the legitimated pathways through academia to be allowed to our students and our colleagues. In short, no American academic with an honest curiosity about the structured structuring structure which defines his or her own activity can fail to have the background knowledge I am about to assume, and academics from other nations will surely understand closely analogous conditions.

Academia, as we in the West know it, is a Liberal institution, at least in its self-image. It prides itself on its openness, in the usual Liberal contrast with the closed intellectual institutions of the past. The claim of openness is part of the core of

Western academia, and it entails, for modern Western academics, the claim of comprehensiveness. The mark of a Liberal academia is its avowed refusal to turn its back on any source of knowledge. In grant proposals we all have to make enthusiastic pronouncements about the social (or other) utility of the research we propose, but we maintain that this simply sets priorities within a limitless range of possible research, all of which is defensible in its own right—just not quite so important as our own. Thus a fundamental structuring of academic possibility space is the resolve to limit its structuring. Just as the first ten amendments to the U.S. Constitution explicitly withdraw certain matters from the realm of normal legislation, the liberal ethos locates control of research in the hands of the individual scholar. But matters cannot be left at the level of ideological pronouncement.

The management of this enormous space of potential research is also meant to be in the hands of the learned professions, the disciplines. Because the individual scholar adopts vows of discipline, the doctrine of academic freedom is the claim to autonomous management by the disciplines themselves. The extent to which self-management is a pose rather than a fact is a question I recognize but will not go into here. What is certain is that the disciplines, through individual departments and through hiring networks, have kept substantial control of entry into the disciplines, and, through that, substantial control of what research is done.

Thus the division of labor within academia is extreme and is becoming more extreme as the internal complexity of each traditional discipline increases. Where it was possible a half a century ago to be relatively *au courant* in all the research areas in a department, the norm these days is that members of a department are unable to evaluate the research of their colleagues except in occasional areas of overlap. This increased specialization is, as we all know, encouraged. It has obvious benefits. It also has less obvious drawbacks, but they are largely discounted. The fractionalization of the disciplines has obvious consequences for the information space available to individual practitioners of any of the academic arts. To mention only the most obvious consequence, the ability of colleagues to learn from one another is diminished just to the extent that the scope of their research contracts. Any contraction of subdisciplines defines an information space (an array, a matrix) of problems considered legitimate. These information spaces become more and more disjoint. It becomes more and more difficult to become conversant with these disjoint spaces. Consequently, should problems arise that are not definable within any cell of the resulting disciplinary grid, it will be less and less likely that the expertise can be marshalled to deal with them adequately. The potential trade-offs here are also obvious. Specialization allows for unperturbed focusing that may be essential at some stages of research.

Specialization at this level is thus analogous to specialization at the organismic level. In both cases there may be "evolutionary benefits." But the drawbacks of specialization in both cases result from a commitment to the stability of the relevant environment over time. In the case of academic specialization survival strategies become skewed toward conservative, active resistance to change of the environment of problems defined within the disciplinary matrix. New "interdisciplinary" ways of conceptualizing research agenda are subjected to resistance *every* time they are proposed, and a condition for their entrance into the research community is that they overcome this resistance. Paradoxically, the neo-Darwinian committed to a linear "pruning" role for natural selection, or an equally linear "adaptation to pre-existing environment" picture must see a big disanalogy where I see an analogy. For the restructuring brought about by the successful entry of research programs outside traditional disciplinary bounds restructures the academic environment for succeeding entries.

Until academia is powerful enough to define the information space for the emergence of problems for society as a whole, the mismatch between the trained intelligence to be deployed, and the problems to be solved will be chronic. Maybe it's better that way—if the only alternative is to run society according to the current academic division of labor. But there is an alternative: to restructure the academic division of labor.

Into this situation have walked a large number of highly motivated women. They have demanded entry into the disciplines, and worked hard to qualify themselves for it. The material conditions for their entry were set by constraints and enablements both endogenous and exogenous to academia. The ideology of Liberalism was always committed to nondiscriminatory participation in political society. In a long struggle women had managed to hold the Liberal *political* institutions to the commitment at least to the extent that the franchise was won. The expanding Western economies, fueled, in particular, by the windfall of the Second World War, had provided entry to the economy for women. Universal education had reduced the time for child care and surveillance required of most women by transferring it to the educational bureaucracy. Academics found large numbers of female students seated in their classes. The academic ethos required that they be treated exactly as the males were treated; and sometimes they were. Many of the female students were very good ones, and they were very often hightly motivated. Liberal academia accepted them, especially insofar as it was incapable, in its own terms, of excluding them.

But of course the structure of academia required that women enter according to the prevailing division of labor. That is, as a condition of entry they had to conform their research and teaching agenda to those legitimated by the pre-existing division of labor. But the same conditions that had encouraged their entry into academia had made them acutely aware of their relationship to these conditions. Although they had succeeded in entering academia, they were aware of the ways in which it was a male academia—dominated by institutional arrangements congenial to male interaction, legitimating questions of interest to males, and populated by males who responded differently to females than to males.

In particular, their perception was that questions directly related to what it is and was to be a man had found a solid place within the curricular division of labor, but no such solid place was provided for questions specifically directed to what it was to be a woman. They have made a convincing case that their perception was correct—largely by showing by their own research the gaps in previous male-dominated research.[3] Furthermore, in line with the view discussed above that the boundary between the knowledge object and the methods of access to it is a difficult and unstable one, women noticed that when the nature of femaleness actually found a place on the academic agenda it did so under the control of research methods and presuppositions that were suspiciously male. Hence the pressure for "Women's Studies."[4]

Now, notice what pressure for Women's Studies is pressure *for*. It is pressure for a legitimated place in the academically constituted division of labor. In contrast, you can imagine pressure to hire scholars within the pre-existing departments and disciplines to teach and do research on the focal subject of women. Structurally, this would have been a redivision of labor of a very different sort, much less threatening to the hegemony of the traditional disciplines. The latter alternative, quite expectedly, was the alternative offered by many faculties who really didn't want to set up an autonomous or semi-autonomous program in Women's Studies.

But what are the crucial signs of legitimation within academe? Among them are, first, having a department, which implies, usually, the right to hire, tenure, and promote (subject to administrative constraints), then having a major, then having a graduate program of your own. The "powerful" disciplines are those that have all of these. Women pressing for a Women's Studies program were asking to be put on the same footing. In other words, they were asking for a restructuring of the institution, and a corresponding share of academic power. At roughly the same time, the same demands were heard for Black Studies, Asian Studies, and, at several institutions, other special studies programs.

As we all know, the timing of pressure for Women's Studies programs was affected greatly by the political climate of the late 1960s. Speculation is possible, but not very fruitful, about the course of a women's movement within academia had the 1960s been different. The fact is that the advocates of Women's Studies found usable space prepared by the civil rights movement and the anti-war movement. In terms of the reform of the academic division of labor, the effort of blacks, women, and other advocates of special studies was a coalitional one.

For academic women, the nature of the effort of the sixties may have been a mixed blessing. On the one hand, the social pressure generated by the civil rights and anti-war movements probably made the difference between winning the fight to establish Women's Studies programs and not winning. It was very apparent that male faculty voted for such programs largely as an expression of their commitment to civil rights clothed as academic liberality. But the complex coalitions entered into by academic women in the sixties had their dangers.

First, where women's studies programs were instituted along with black studies, etc., acute problems of organization and consolidation arose simultaneously for all the new programs. Where departmental status was not won, faculty for the programs had to be recruited largely from the existing pool within departments, so turf and workload issues arose. The new programs inevitably ended up competing for faculty—not only with each other but with the "home" departments. Since the departments retained ultimate control over hiring and firing, there were immediate disagreements about hiring and staffing priorities. Under these structural pressures the weakness of the coalition became obvious. Academia could, in effect, isolate the coalition members from each other and convert them, again structurally, into competing self-interested parties within a system of scarce resources. As usual, the scarcity of resources was highly manipulable. The availability or nonavailability of resources within academic institutions is notoriously sensitive to the craft of creative accounting. But, in addition, at many campuses the fear that the college-age population was going to get smaller, coupled (in the public and semi-public institutions) with the reluctance of legislative bodies to pay for inflation-related increased costs of education, made the ability to attract students a *de facto* necessity.

It is sometimes noticed, but not often with enough clarity, that the primary method of assimilating academia to market society is the commodification of students. (The other main method is the commodification of research.) Furthermore, mainstream academia is very docile in its acceptance of this commodification. Pious noises are made about the sanctity of academe and the unsuitability of commodifying knowledge, but in fact these noises are not translated into action, for to do so would be to adopt a radically negative stance to the dominant form of social structuring. Academia depends, in these days, too much on the good will of the dominant culture to press its claim to special status in this regard. It has to appear responsive to job market pressures, supply what the economy demands.

With the commodification of students in place, the nascent Women's Studies programs faced a very clear set of structural parameters. They either filled their classrooms or they lost their ability to command academic resources. Departments could argue that released time for individual faculty was unprofitable. They could argue that hiring priorities ought to reflect student demand.

One would have expected a socially sophisticated (largely radical) faculty to apply itself to dissecting the structural constraints on student demand to show how the supermarket model of academia was acting to channel research and study opportunities. The only attempt even to begin to do this was a caterwauling about the vocational preoccupation of students. (As if the students had real alternatives and were somehow "to blame" for the shape of the possibility space *they* had to confront after graduation.) And with the problem conceived in this way, there was the overwhelming temptation to deal with the distribution of students across courses by means of the inevitable modification of curricular requirements. But then, of course, why should large numbers of students be required to take Women's Studies courses? The long-embedded structure of the established disciplines was virtually unmovable.

Another important aspect of the rise of Women's Studies programs and their coalitional affiliations is that they succeeded (insofar as they did) in the sixties: For, internal to the women's movement both then and now is a spectrum of opinion which must be mapped on the axes of traditional social thought. There are and were moderate feminists, radical feminists, explicitly Marxist feminists, and liberal feminists who conceived the issues in terms of civil rights, etc. An exactly parallel spectrum was to be found in the Black movement, and yet another in the anti-war movement. So the coalitions were all at least two-dimensional: one dimension deployed along the axis feminist, black, anti-war; the other along the traditional axis drawn between the poles of radicalism and conservatism. Given the possibility space of the American sixties, it is fairly clear that success depended upon the ability to hold together these spectra to display a united front. This succeeded at many institutions at least to the extent of producing Women's and Black Studies programs, and the denial of credit to ROTC courses.

But then the time came when all these internal differences had to be managed within the space defined by the commodification of students and the competition for their presence in the classroom. I am not being snide when I say that every institution wanting to retain and stabilize "special studies" programs needed a strong black woman with leftist sympathies. And I am only being a little snide when I say that it also helped if she were gay. *Some cohesive presence had to be established.* But the differences internal to the forced coalition were not just "differences of opinion" (and these are the only differences liberal academia knows how to manage); they were real differences within somewhat unstable communities. In particular, the question of women and radicalism within the black community is notoriously difficult (often conceptualized in terms of the power of the storefront baptist churches versus that of young males). This issue (along with many other similar ones) was imported inside the walls of academe. It was fought out in the battle for establishment and inital control; it was fought out in the battle for students.

Determinism Revisited

From the vantage point of extreme abstraction, the preceding story is one of race and gender. Furthermore, the abstractions—gender and race—are ones of prime biological relevance—for something or other. The question is if, and how, they are relevant to explanations of the rise of Women's Studies programs.

One strategy here would simply be to challenge a sociobologist to reduce all the explanatorily relevant conditions, or, more realistically, to provide a reasonably detailed program for the reduction. This is a standard strategy in the polemical literature directed against sociobiology. But I will not do this, for two reasons. First, such a challenge comes very close to conceding the conception of explanation I have been contesting, merely adding that the sociobiologist cannot provide such an explanation. But in these terms the alleged impossibility can never be made to stick, since the sociobiolgist can always lay off temporary failure against future success. And second, I did not go to the trouble of providing a detailed account just to throw it in the sociobiologist's face, which would be a pretty cheap trick.

A far more profitable approach is to remind ourselves of the lessons of robustness and multiple access. For, as we look at the complex of constraints within which the emergence of Women's Studies programs takes place, it becomes apparent that no one investigative path—as they are normally demarcated—is sufficient to unearth all the relevant structured structuring structures. Furthermore, since we cannot assume additivity, a piecemeal and isolated one-by-one investigation of the relevant facets of the situation cannot be presupposed to yield the understanding we want.[5] The various investigative pathways are as interpenetrating as the structures they investigate. Of course we want to avoid thinking of this as the start of a radical distinction between the natural and social sciences. The very same investigative necessities exist for natural phenomena, as I tried to show earlier. In these days when the developmentalists, population geneticists, and molecular geneticists find it unproductive to isolate themselves from one another, a point about robustness and multiple access is much easier to make than it would have been even a few years ago.[6]

A biological determinism, that is, a view that possibility space is totally determined by biological factors, must be held in mind as a sort of default view, against which I am trying to pit the dialectical dynamics I have developed here. The example we just looked at involved a highly structured practice, the academic practice, which is on the face of it the multi-dimensional consequence of a long complex history. A biological determinism would have to articulate this history and all the resulting dimensions in terms of biological determinants.

Biological determinisms seem constantly to be in search of human species universals. While this search is motivated in the first instance by the search for explanation as derivation, it is motivated in the second instance by the following conception: What is genetically universal in human beings will always be expressed in human behavior and institutions. If this is the claim that everything in the genome will be expressed, it is known to be false for virtually every organism studied with any care up to the present time (with the possible exception of the viruses). The general picture, especially for eucaryotic organisms, is that some of the information in the genome is expressed and some is not. *Whether* such information is expressed, and *how* it is expressed depends on a complexly organized regulatory system operating at molecular, cellular, and organismic levels. In fact, the system of constraints can act in such a way as to make certain genomic information, universal in a species, utterly irrelevant to a phenotypic account of the species. Or, more generally, the information is systematically integrated with other genomic information so that its "expression" is radically different from what you would expect on the basis of the original information itself.

The pull of metaphysical rationalism is very strong in all determinisms. Correspondingly, determinisms trade on some version of the thesis that explanation is derivation. The positivist version familiar to us all is simply the latest in the line of such theories winding its way back to Pythagoras and Parmenides. If biology *must* be thought of as a quasi-formal system—because all intellectually responsible fields of investigation must be thought of in that way—then, yes, explanation must be derivation, and failure of deducibility constitutes failure of explanation. But deducibility requires universal major permises; hence the search for the universalities required. And if sociobiology is to take its place alongside the other rationalist enterprises, it too must find its legitimation in its ability to provide the human universals required as major premises.

There is no *a priori* reason to believe that the constraints generated by historically evolving social systems do not themselves integrate human "species universals" in ways that either suppress or modify their expression in human behavior and institutions. Thus the task of a sociobiologist has begun, not ended, when he has plausibly identified such a "universal," for he must then investigate the modes of expression and the possibilities of nonexpression of the universal. Only when he has shown that the universal *must* be expressed in a canonical way in *every* circumstance will the strong thesis of genetic determinism be demonstrated. There is, however, a weaker thesis available: the "default drive" thesis we saw earlier. This is the claim that human universals would indeed drive human action were it not for the social constraints that mask its effects. The real difficulty with the default drive thesis in a *scientific* study of society is that it is utterly immune from experimental test because, for example, any experiment would (in addition to being pretty horrible) simply impose a bizarre *new* set of social constraints—namely, the experimental constraints. This renders the default drive thesis absolutely irrelevant to the explanation of human action. The problem is one of closure, related to the problems of inference from laboratory to nature examined in chapter 4.[7]

From the rationalist persepective, then, the "theory" of explanation developed over the course of this book is seemingly inadequate. What I say is, and is meant to be, utterly unconvincing when measured against the baseline of skepticism and the search for an omniscient epistemological perch it implies. For the view of sciences as quasi-formal systems is explicitly intended to answer to the skeptic. On this view, scientific truths have to be quasi-theorems. But derivability is a forlorn hope. Or, to put the matter in terms already introduced, the relation between an explanation and what it explains is indeed a relation—not a function. Embedded in fields of structured structuring structures determining their possibility space, events are seldom derivable with great specificity from the conditions from which they stem. Every scientific theory seeming to allow derivability, and this is to say every scientific theory amenable to axiomatization, is a theory of and for abstractions. It simplifies in terms of a set of abstractions specifying a small number of parameters for the relevant phase spaces; it is applicable only when the parameters define just those degrees of freedom adequate to explain a *facet* of the behavior of a complex phenomenon that happens to be the one we are interested in. Indeed, if I fall off the top of a tall building, thus converting myself, for all intents and purposes, into a ballistic object, Newtonian mechanics by itself will do quite nicely to predict my immediate future.[8]

My own view, following Bourdieu, is that through both biological and social history, structured structuring structures are generated in such a way that we can always give an account of their genesis in terms of the material conditions out of which they emerged, but that at any given stage it is impossible to reduce the resultant phenomena synchronically to any one dimension—be it the physical, biological, or economic. In addition, within the complex interactive systems generated, it is possible that some of the dimensions are modified significantly in their importance and activity—even to the point where they may "drop out" of adequate contemporary accounts. For example, it could be that many of the important aspects of being a woman, aspects crucial to historical dynamics in the past, may not be particularly relevant to explaining gender-centered phenomena now. The material conditions for sexism contain among them our ability to tell one sex from another. But for a clothes-wearing species in which sexual dimorphism is minimal within a large area of overlap, even this apparently basic condition can be rendered irrelevant in numerous particular cases.[9] Prior structured structurings may have organized such aspects so as to absorb them institutionally and render them locally irrelevant. To think otherwise, according to my view, would be to ignore history and to conceive of possibility space as a timeless void of equi-possibility. Liberal rationalism conceives matters in this way. Not I. In any case, the failure of explanation as derivation should not be thought of as marking the boundary between the "natural" and the "social" sciences. As we have seen in several ways, the failure goes far deeper than that.

Teleology, Function, and the Pseudo-Heracleitus

Finally, I would like to stand back and assess the merits and limitations of the account of the rise of Women's Studies I have used here as a vehicle. For example, can we expect a tight deterministic account of a general phenomenon such as this? Maybe not. It may not be until we get to unique particularity that all the pieces move into place and we get the sort of explanatory closure we dream of. That is, it may be that all the structured structuring structures we can dig out, all the material conditions we can identify, are insufficient to lock us in deterministically. Remember, structured restructuring is an enabling as well as a constraining process. So what looked like closure under a previous structuring may not be closure under new material conditions. This, for example, is one of the lessons of our discussion of Malthusian closure a few chapters back. Despite the fact that all phenomena occur under thermodynamic constraints, and all biological and social phenomena occur under biological constraints, the possibility space they jointly determine may not be tightened down to a single path.

Several pathways may remain open as possible futures for things or persons when all the structured structuring structures are accounted for. In these cases we would often want to know why a given thing or person went down one path rather than another and would quite naturally be led to look for the pushes of efficient causation or the pulls of final causation. The result would be an explanatory pluralism. Such a pluralism is in any case explicit once a Garfinkelian framework is adopted. Pluralism of this sort should bother us only if we have a metaphysical commitment to a one-dimensional determinism. It helps in this to have the view of physical laws I laid out in chapter 2. So, in the case at hand, if someone were to say that an explanation of the institutionalization of Women's Studies at a particular college was incomplete without an explanatory component relating individual aims to social consequences, our pluralism could easily allow us to agree. In fact, it would be pretty silly not to, so far as I can see. But we might be wise to retain one sort of disagreement, and one related failure of interest. First the possible disagreement.

We might disagree with the advocate of explanations in terms of individual aims. I would insist that there are structuring constraints determining both the "individual" and the "aims." Today's women *can have* a conception of themselves that women in earlier times *could not have had,* and vice versa. The individuals who act, therefore, are not dehistoricized "agents," but hightly determined real people. Similarly, the aims academic women *can have* constitute a highly determined possibility space, a space very different from those of earlier women. Thus my failure of interest in the voluntaristic part of the explanation. For it seems to me that the job of a social science is to figure out how such individuals and such aims are possible; and the major part of *that* job is the articulation of the constraining and enabling structured structuring structures. Bourdieu, at this point, would insist that many of the "choices," "decisions," and "resolves" of academic women were far more tightly constrained than the women themselves would like to think—that many of the choices made were "virtues made out of necessity." So for him, the realm of freely chosen purposeful behavior in the situation would be quite small. To decide who is right about this would require a full analysis within a well-established social scientific framework. We may or may not have such a framework at our disposal. In any case, the explanatory view I have offered can accommodate purposive human action but tends to restrict its range far more than the usual Liberal theories of action. Issues about this range are likely to turn on efforts to specify the "individuals" and on the stringency with which their possibility space is determined.

Another challenging set of problems arises with the attempt to move to teleological explanations not involving human agents—that is, with attempts to say that things happen "for the sake of" some end, or that the "function" of something is the attainment of some end. On some views, indeed (Campbell 1985), these functional explanations are virtually definitive of biological process and/or human societies. It is partially to address these views that the current chapter is couched in terms of the division of labor. For it seems natural to ask why the academic division of labor exists and what the effects might be of modifying it by introducing "cross-disciplinary" activities such as Women's Studies. These questions lead just as naturally to the question of what purpose the academic division of labor serves—what function it performs. These questions look very like parallel questions about opposable thumbs, appendices, and air bladders, questions that arise in a physiology focused on questions of structures and their functions.

In keeping with what I have said before (e.g. in chapter 4), I see no reason why talk of function cannot be explanatorily useful, *so long as the requisite closure conditions are respected.* In physiology, the requisite closure conditions seem so obvious as to need no mention. The healthy functioning of the organism, or the "normal" functioning of the organism is naturally adopted as the condition of teleological closure. Once it *is* adopted, then talk of function should cause no problems (even to the strict mechanist, who can point to the closure condition as nothing more than a license for the poetical talk of function). We do, however, have to be careful about where the teleology lies. And here Garfinkel's insistence that we know what question we are answering as we offer an explanation has to be taken scrupulously to heart.

"In order to pump blood" is *not* an answer to the question "Why does the heart beat?" unless the teleological closure condition has been invoked (albeit tacitly) as the determinant of explanatory space. Similarly, though perhaps less obviously, the answer, "Because of intermittent electrical pulses from the brain stem" to the same question is only an answer under mechanistic closure conditions. Thus *both* the teleology *and* the mechanism are highly truncated with respect to the totalizing aspirations of their adherents. The mechanist is barred from saying "It's only a machine" unless he can show that the mechanistic closure conditions *must* be invoked to the exclusion of all other explanatory closures. Correspondingly, the teleologist is barred from attributing goal-oriented behavior to the heart. The closure conditions that license the functional explanation do nothing more than set stability boundaries on the organismic system and locate the place of the heart within them. Perhaps the resulting modest sense of function will satisfy the teleological urge (it satisfies mine); perhaps it will not. For it falls far short of implying any grand design.

Similar points can be made about the function of the academic division of labor, but not so easily. For one thing, there may not be any easily agreed upon closure condition. There may be disagreements about the healthy or normal functioning of academia. Its role in the broader social context may be disputed. Academia is complex enough to have a possible multiplicity of "functions" itself, so the functions of its internal organization may be correspondingly multiple. Given the multiple functions of the institution itself, it may not even be possible to assign functions to particular internal structures without uncovering incompatibilities. Some structurings may facilitate the performance of some academic roles and impede others.

The rise of Women's Studies may in fact have smoked out some of the potential incompatibilities within academia. From the point of view of the bureaucratization of the professions, and their corresponding capacity for self-regulation, the department structure is a highly adapted organ; but from the point of view of the Liberal ideal of free inquiry it can often be an impediment. Similarly, departmental structure, and the corresponding regularization of "majors," is highly efficient in performing a clear-cut credentialization job. Consumers of the academic product know what they are buying when they hire a credentialized graduate. The proliferation of "cross disciplinary" studies would muddy the waters of credentialization—be "dysfunctional" with respect to the orderly transition to the job market.

Of course, the bureaucratization of any institution, even academe, also brings with it the possibility that the institution will lapse into a haven for the bureaucrats, and that modifications of the internal structure will be made for their security and well-being. This further complicates any attempt to assign function to structure.

So within a highly complex social entity such as academia, finding the appropriate closure condition to generate functional explanations for aspects of its internal organization is likely to be a chancy and fragmentary affair. Clear-cut teleology is difficult to argue for in these cases, despite the fact that the internal structure of academia is *thought of* as under the "rational" control of some very intelligent people, whose aims are supposed to be pellucid. To move to less rationalized social subsystems with the hope of finding functional explanations would seem on these grounds to be very risky. The voluminous literature discussing latent and manifest function, and the difficulties which have arisen in that literature, only scratch the surface of the problem. Every functional explanation needs its closure conditions; and for multifunctional structures the system of closure conditions need not be expected to be consistent.

There are additional, perhaps deeper, difficulties with functional explanations, and, indeed, "efficient cause" explanations. The conditions enforcing explanatory closure for a functional explanation set static explanatory baselines. Very often, as with the assignment of function to organismic structures, the static baseline is the assumption of equilibrium—homeostasis. But from an important point of view organisms are not in equilibrium, but far from it. The evidence, say, in mammals that they maintain an "equilibrium" temperature when they are healthy is thus very misleading, since they maintain themselves in this state not by staying in equilibrium with their environment, but by extracting energy from it in one form and dissipating in another—by eating and basking, and by giving off heat and other waste. The choice of an equilibrium state as the baseline for functional explanatory closure, then, may be all wrong, or so partial as to be dangerously misleading. When we move from organisms to social systems the problem may be even more debilitating. The next chapter attempts to begin to deal with this issue in some particular cases.

CHAPTER 9

The Formative Dynamics
of Cities

One of the major problems in constructing genuinely dynamic theories is the overwhelming prevalence—both in biology and in social theory—of models oriented toward equilibrium. The advantages of these models are obvious. They establish "inertial" baselines against which the effects of exogenous forces can be assessed; and they provide the conditions under which equalities can be found so that systems of equations can be solved. Unfortunately, as helpful as equilibrium models can be, they are also the creators of illusion. Until recently it was very difficult to appreciate this fact—let alone to do anything about it. But now, thanks to the work of Prigogine and others who have exploited his results, we have the foundations of a model of open systems to pit against the older equilibrium models. For we now can talk of open systems stabilized far from equilibrium, and ask of familiar systems whether they are such systems. The study of these systems is called nonequilibrium thermodynamics (henceforward NET). Up to now NET has been used primarily to study chemical systems and ecosystems.

Recently there have been several interesting attempts to extend the methods and models of NET to economic systems, approaching them as particular sorts of ecosystems.[1] The results seem to promise that such extensions will prove useful. So far the attempts have been made by people more familiar with the biology of ecosystems than with the economics of social systems. Consequently, the style and language of the most interesting work may not yet allow social scientists to appreciate the potential of NET models. Here I will lay out an account of cities as dissipative structures which makes more direct contact with the language and concerns of economic theory. My intent is to offer a specimen of the exploration of NET techniques that is sufficiently plausible to tempt further explorations by others. My discussion will be far more qualitative than quantitative, despite my view that we have to move quite quickly to the point of seeking analytical results.

At the end I will discuss some of the methodological issues central to achieving such results.

Dissipative Structures

For those not yet familiar with the concept, dissipative structures are "spontaneously" arising stable configurations within an energy flux. The best studied of these systems are vortical structures arising under specific convection conditions in relatively uncomplicated chemical systems (Lugt 1985). The most "familiar" are meterological systems such as tornados and hurricanes, although this familiarity is an artifact of their relative simplicity. From another perspective, *you* are the dissipative structure most familiar to you. Dissipative structures exhibit order and stability. They maintain their internal order by utilizing an ambient energy flux and dissipating degraded forms of that energy to their environment. Internally, their order is maintained (or even increased), which occasions an "entropy debt" that is paid by the increased disorder of their environment.

Let's look at dissipative structures explicitly in terms of the second law of therodynamics. We want to track the entries that an accountant writes in a book as he takes measurements of a process at intervals of time. The second law in its various formulations says that these entries will be related in a particular way—in an increasing sequence with the possibility that two successive entries may be equal. The classical formulation says that the accountant is to look at the dial of a calorimeter; Boltzman says that he is to look for patterns and discontinuities (and that such patterns and discontinuities will become less frequent and finally disappear); and Shannon asks the bookkeeper to write down how many marks he has to make in order to tell the state he is looking at from any other (whereupon the second law would demand that it takes fewer and fewer marks), but despite the similarities it is very doubtful that the second law holds for Shannon entropies (Wicken 1987).

What Prigogine did was to take the standard equation for entropy production and divide the entropy term into two parts: roughly the entropy resulting from processes internal to a system, and that resulting from its interaction with its environment. That is, he tells the bookkeeper to look at two different dials. Then, it turns out that the readings on the "internal" dial need not behave in a second-theorem way. The Prigoginian equation tells the bookkeeper how to combine the results of the two readings and that the combination conforms to second theorem requirements.

Furthermore, the way in which the internal readings can change is related to corresponding changes in the external readings. The curve depicting the relation is different for different systems, meaning that different systems have different efficiencies in exploiting their "environments." But once the curve is found (based on Gibbs' equation) we can speculate about what the bookkeeper is going to find many entries down the line. When the system is near equilibrium, or at it, we can always tell what the next entry will be, given the curve and the preceding entry. But far from equilibrium we can get to a point where we cannot tell which of two (or more) entries will appear. This is a *bifurcation point*. It has nothing whatso-ever to do with the system splitting into two parts, but, rather, with the relevant equation having two equipossible solutions.

As an analogy, think of going out walking and coming to a narrow gate in a wall. You cannot get through frontways, so you have to turn and go through side-ways. Your private accountant has been following you, keeping track of your ori-entation. He has been saying "front," "front," . . .; and now he will have to say "leftside first" or "rightside first," but there is no way to predict which he will have to say. He does know, however, that you will not divide in two to get through the gate.

That is, turning leftways and turning rightways are two stable solutions to your walking curve; splitting in half is not. (This is a consequence of your particular structured structuring structure. Something else *might* split in half under the same circumstances.) Similarly, curling up one way or curling up another way may be stable solutions to the equation for something else; and we may not be able to know beforehand which way it is going to curl.

Now, if the slot in the gate were a tunnel, and you started in leftways, you would stay leftways for the length of the tunnel. The constraints would not allow you to turn rightways. The conditions of flux constrain you to the single same regime. Or, if you kept coming to gate after gate with only a little free space in between, it might work out that you did not have time to turn frontways between gates; so unless you stopped between gates (analogous here to dying or falling apart) you would always go through gates the way you started going through them. This is analogous to a periodic flux situation (Johnson 1981).

Notice how little "causal" language there is here; how little temptation to say what "drives" your leftways or rightways behavior. Now suppose you were walk-ing and kept coming to gates. There is enough room in between to turn, but sup-pose turning were really exhausting; those who kept turning frontways then side-ways soon fell down and died. Then, to keep going, you would have to walk crabwise, gate or no; and the first way you turned would be the way you ended up. This means that the given energy budget and the presence of gates *determines*, but not Humeanly, your walking attitude. Once you pop into a stable regime (left-ways, for example) you stay there or drop out. For a realistic example, think of walking into the wind. The general point is that at a bifurcation point a radical discontinuity must occur in order for the system reaching that point to persist after it.

It is clear enough that the dissipative structure has to maintain phase separa-tion from the flux—must develop an inside and an outside so that you can iden-tify differential rates of internal and external processes[2]—so must be complex enough to maintain the boundaries. And as flux changes take place, the require-ments on the ability to exploit the available energy may become more stringent; and there thus may be "selection pressure" for increased efficiency. The seductive analogy here is the division of labor, i.e., the Adam Smithian claim that a system of linked specialists is more efficient than a group of autonomous jacks of all trades. And, of course, ecological systems have a life cycle that involves the replacement of initial populations of generalists with an organized system of specialists. But notice what is going on here. The linked specialists could be simpler than their predecessors. The increase in complexity is in the system of systems; not in the system you started with, which could well become simpler (one-dimensionalized, for example). So you have to be sure you know the locus of complexity. Ecosystems become more complex; mutualisms become more complex; organisms become more complex; but constituents of any complexification can well become simpler. Some people think that viruses are simplified versions of previously more complex organisms. Their reproductive life is complex, but that is another matter (or another mater).

NET generates models of evolutionary processes which contain both "initial conditions" and "boundary conditions" components. Systems with one sort of internal order will respond to boundary conditions in one way; systems with *another* internal order will respond another way. No explanation of the response is complete without reference to the internal order—the structured structuring structure.

Imagine we have been able to identify an ecosystem as a suitable dissipative structure. Then we ask if it is reasonable to consider it as an *isolated* system, that is, one where neither matter nor energy enters or leaves. But as an isolated system it would collapse into equilibrium rather quickly. We remember that there are already decisive reasons for thinking of it as a far-from-equilibrium system, and this precludes its isolation. So we think of it as a system where energy goes in and out. Then we ask about the internal dynamics of the system: the dynamics that are solely due to the internal organization of the system, independent of energy input. We are asking about the shape of the possibility space and the way it changes, assuming that all the changes are due to internal events. (We mark the use of *due to* as a potential puzzle.)

A pure initial conditions account of an ecosystem would by itself answer all questions about its evolutionary dynamics. That is, we would have to ask how the ecosystem would evolve given only the trophic relations, the mutualisms, etc. But in this case we would soon give up a *pure* initial conditions account as a puzzlingly unreal abstraction. Just to mention one reason why, we would soon see that playing out of the internal dynamics of the ecosystem changes the boundaries of the system. Thus initial conditions and boundary conditions are nonindependent, so a pure initial conditions account is not available here. *But* neither is a pure boundary conditions account.

Now we go to a species[3] as a possible dissipative structure, knowing that if a pure initial conditions account of its dynamics is to be possible, then initial conditions and boundary conditions have to be independent. (This is parallel, by the way, to decoupling variation and selection.) The relevant assertion is, "The information content of the species does not change the boundary conditions within which internal changes must occur." An immediate inference from this is that no internal genetic change can make you tastier. If it does, it has changed the boundary conditions.

I would suggest again here that no pure initial conditions account was possible for either system. But to conclude that a pure *boundary* conditions account was possible seems to me equally bizarre. It would involve the claim that the initial conditions were irrelevant to future dynamics. Given my view of the importance of initial conditions as the material conditions determining the possibility space of future dynamics, I could hardly go for a return to boundary conditions models. Prigogine is in agreement on this, as is Salthe. In fact, dissipative structures as Prigogine defines them require for an account of their dynamics both initial conditions and boundary conditions to be specified. There is no reason for this to change when you move into the realm of organismic evolution. (See chapter VI or *Order out of Chaos*.)[4]

As we know from reading Prigogine and others, one of the major tasks in dealing with NET phenomena is setting up the criteria for the systems to be dealt with: their boundaries and the background conditions within which they arise and are sustained. Any dissipative structure consists of a system with enough internal coherence to utilize a background flow of matter and energy to *sustain, maintain, and reproduce* this very internal coherence. As the work of Prigogine, Wicken, and others suggests, this internal coherence, under the right conditions, can even become more elaborate and differentiated over the course of the history of the dissipative structure. So we know that if human institutions and systems are dissipative structures we have to identify structures and relate their ability to sustain, maintain, and reproduce themselves within the limits of the resources (material and energy) available to them. These resources, so to speak, flow by and through the systems, and the successful systems find ways of organizing themselves to be able to utilize the available flow. The first question is, "What are the appropriate systems to look at?" In the modern world we seem to have a lot of choices, from families to nation states. From an economic point of view, families, villages, cities, regions, provinces, states, etc., all the way up to nations, seem to be prospective candidates for investigation as dissipative structures. Furthermore, economic accounts are kept at each of these levels; and decisions are made on the basis of calculations of the future behavior of these accounts.

Such thinking quite naturally leads us to think of the increased interdependence and differentiation of institutions as a consequence of what interdependence and differentiation contribute to the sustenance of the whole. A familiar story. But each increment of interdependence requires new devices for establishing the internal coherence of the larger system. And each new device has its cost, its "entropy debt." The larger system may indeed be able to balance its internal accounts in the manner of economic equilibrium theory, but the mere balancing of the accounts does not tell us anything about the relative costs of sustaining the new larger system as opposed to the old smaller ones, or where these costs are paid.

What *NET* does, then, is show us that there are interrelations between the social structures we have and the rate of material flow required to sustain them.[5] Or, in other words, it tells us that the entropy debt incurred by our elaborate organization can be paid in several ways: that the information content of our social system is necessarily connected to the rate of material flow needed to sustain it. For example, just to fix ideas, the standard line (since Hume) has it that economic systems have moderate scarcity as their condition. But scarcity is not the primary condition for an economy. What economies rest on are gradients. They depend on finding ways of keeping material flow at a suitable rate. Sometimes this is recognized by economists themselves, as in the concept of comparative advantage in discussions of international trade. More often, however, the need for gradients is *misrecognized.*

Jacoban Cities

At this stage we need something a bit more concrete to work with. From a number of sources I could have chosen, I chose to discuss a theme from Jane Jacobs' *Cities and the Wealth of Nations* (Jacobs 1984). Her analysis in this book and the earlier *The Economy of Cities* (Jacobs 1969) raises very convenient questions for my present purposes. In addition, it gives us a clear set of structures and issues to work with before we are vexed by the problems of measure and measurement which eventually have to be confronted by NET.

Jacobs argues that despite the bifurcated emphasis on individuals and nations as the important constituents of economic systems (in micro and macroeconomics respectively), the dominant structural components of economic life are import-replacing cities and their associated regions. She gives no general account of the rise of such cities. Perhaps there *is* no general account. But once in place these cities organize an economy that prospers over a significantly long run. These import-replacing cities are cities that manage to become increasingly self-sufficient partially by modifying the nature of their import requirements. They are to be contrasted with other cities which *do not* organize viable economies. Economic life in and around them is very different from that around the favored cities.

Now, anyone reading Jacobs' book who is acquainted with NET will surely note the resemblance of these favored cities to dissipative structures. The lack of a general account of their genesis is itself indicative of the failure of linear causal models to account for them. The import-replacing cities look for all the world like strange attractors—structures that emerge from economic flux, and then begin to organize that flux. This model, of course, is consistent with a wide range of explanations for why a favored city arose here or there, at this time or that. But the explanation need not be "economic" in the same sense that would be appropriate to an account of its continuing stability once established.

We may be able to see this in terms of one of the dominant themes in NET, namely the recurring theme that a stable structure far from equilibrium must succeed in altering flux rates sufficiently to create clear-cut phase separation. That is, internal order must be created which is far more efficient in utilizing energy for organization and maintenance than the background system within which the primary flux occurs. (Lionel Johnson's study of the subarctic lakes can serve as a reference point here [Johnson 1981].)

When we weave this theme into Jacobs' analysis we get some illuminating results. First of all, we are led to notice that there are two importantly different sorts of trade, where the dominant theories of economic analysis give us only one. In one sort of trade the mutual flow of material and energy simply tracks gradients. The trading *process* is simply a gate through which flow takes place. *This* sort of trading does indeed tend to lead to equilibrium—in the classic Boltzmanian sense. The redistribution resulting from trade tends to eliminate the gradients. Depending on the time-dependent fate of the traded material, equilibrium may actually be reached, approached, or never reached.

Examples of this sort of trade are extremely common, of course. Historically there are even some conspicuous cases of this sort of trade persisting virtually by itself for relatively long periods. The Islamic traders both in Africa and on the Indian Ocean provided the gates for such trade. Much of the trade of Venice and Genoa *in their early days* was of this sort. The Dutch replaced local merchants as the gatekeepers of such trade in the Malaysian Archipelago (Braudel 1984, Cipolla 1980).

As we think of examples of gradient-tracking trade, we are bound to notice that in some favorable circumstances stable identifiable structures emerged, fueled by this trade. These identifiable structures can initially and typically be called markets, if we remain prepared to accommodate future complexities which convert them into fully formed economies. When I talk of markets as identifiable structures I do not mean simply that phenomena arise susceptible to bookkeeping of a certain sort. This is the trap which must be avoided. Nearly every human activity (and many non-human ones) can be squashed into the Procrustean bed of orthodox economic bookkeeping, obscuring nearly every feature of institutions and practices we need to know to get a satisfactory account of their historical dynamics. When I talk of markets, I mean established spatio-temporal structures organized within the flux of gradient-tracking trade.

Not surprisingly, these structures grow up at the gates. They start out as stable, reliable managers of the gradient-tracking flux. But in favorable circumstances these market centers begin to use their advantage as gatekeepers to divert part of the flux to the creation and elaboration of their own internal organization. The part of the flux they internalize fuels an internal division of labor and a consequent differentiation of social function, which allows space and time to be cleared for the activities that eventually result in what we would recognize as developed civilization and culture. In other words, the coalescence of these structures—first markets, then economies—results in just what we would expect on the basis of NET, the development of an internal economy. With this development there arises a second sort of trade: organization-promoting trade.

Organization-promoting trade is the allocation and distribution of material and energy (including human energy) *within* an economy. It is used to produce and reproduce the structures necessary to achieve the efficiencies in energy and materials utilization which result in, and are promoted by, phase separation.

Now that we have both sorts of trade preliminarily in view, let's pause for a moment to examine their relation at a next stage of clarity. I mentioned above that the gradients necessary for economic activity are often misrecognized. We can now see better why this is so. Every economic system sophisticated enough to be available for analysis is already fairly complex, and generally (but not always) has adopted a medium of exchange which tends to obscure the distinction between the two sorts of trade. In other words, the accounts tend to be kept in a uniform accounting system. This is what, in the end, allows us to keep *national* accounts, despite the fact that, as Jacobs points out, these accounts are tremendously misleading as a picture of the health or sickness of the regional economies added together in a national lump. Given the uniform accounting system, it is hard to see that there are two different sorts of trade at work.

But we have to remember Jacobs' key concept of import replacement. For it can be understood precisely in terms of converting gradient-tracking trade into organization-promoting trade. Part of the gradient-trade material and energy is diverted to internal organization. This, in fact, is what distinguishes an economy from a market. The initial organization of a market comes about through the establishment of a determinate place for a gradient-tracking trade gate. In addition, the market becomes organized according to rules of access and trading procedure. At a next stage, systems of credit can develop—the first internal division of labor and function. The crucial change from market to economy takes place when those who are associated in the market organize themselves in such a way that they can change the flux qualitatively. What used to be imported is now manufactured internally, for example, and the imports tend to consist of more primary material. The nascent economy is now differentiated at least into a manufacturing and mercantile sector. But soon, under favorable conditions, the differentiation becomes very much more complex.

Notice that the distinction between gradient-tracking trade and organization-promoting trade does not reduce precisely to the distinction between external trade and internal circulation. The reduction is blocked when we recognize the ways in which organization-promoting trade generates and maintains structure. This structure generation may continue to be lacking within a social system which has been unable to constitute itself as an economy. Jacobs gives a number of examples of such social systems; many more are available in the broader historical accounts provided by Braudel and others. Furthermore, economies are normally constituted by subsystems of various sorts, and some trade internal to the economy as a whole is gradient-tracking with respect to the subsystems. Last, it ought to be obvious that organization-promoting trade is gradient-creating. It generates new differential needs.

Not all flux-promoting gates become markets. Not all markets become economies. Just so, of course, not all dissipative structures arising out of chemical flux achieve a degree of internal coherence that allows them to become lasting and stable phase separated systems. This leads to the obvious question of why some evolutionary pathways succeed and some fail. Theories of biological evolution are, of course, largely concerned with this question. In the economic sphere, some of the following considerations are useful to keep in mind.

Geographical location is a crucial variable in the rise of economies as dissipative structures. But no one set of rules will define the nature of the geographic variable for all time. The development of economies generates new efficiencies in the management of energy and material flow. Some of these efficiencies are explicitly transportational and/or concerned with information management and dispersal. Geographic variables are importantly transportational and informational variables, so developments in transportation and communication redefine the geographic base of viable economic structures.

Second, economic structures are interpenetrated by other social structures—e.g. political and military structures—which can have a semi-autonomous history with respect to any given economy. For example, a military capability may develop within the marauding semi-nomadic way of life of an essentially rootless people. However, if this capacity is attracted and trapped by a market, it may function as one of the material conditions needed for the market to develop into a fully articulated economy. Military capacity can temporarily seal off the economy from buffeting flux and allow phase separation to be secured within the system the military strength has "artificially" sealed off. Eventually, of course, the internal structure of the system must become efficient enough to provide the surplus upon which the military depends. (Alternatively, the military can return periodically to marauding.) In the modern world, when an economy is weak and incapable of sustaining its own organization, the military and police power, fueled from outside the economy, become a permanent, dominant feature of life.

Finally, we can think of an extreme situation ecologists often encounter: a species adapting itself to one dimension of its environment, rigidly committing itself to maximal efficiency within that environment at the expense of the capacity to react effectively to environmental change. Such species are always in grave danger of being evolutionary dead-ends, because environmental change destroys the narrowly circumscribed conditions for their survival. (Notice, by the way, how wrong it would be to attribute this process entirely to "natural selection." Without an account of the material conditions embedded in the ancestral genome, including its ability to generate variability, the natural selection explanation would be totally adventitious—truly tautological.)

Similarly, we can think of the "one-product" city: for instance the mining town. It is extremely vulnerable to changes in, let's say, the economic climate. Organized rigidly for one purpose as it is, if the lode dries up, or if the market changes, it will take enormous effort to reorganize for success in the new environment. *Here is where the entropy debt is paid!* This may be called the mining town/ghost town phenomenon. Ghost towns, always single-resource towns, find it impossible to marshall the resources to reorganize their economies for success after significant changes in the economic climate for their single resource: the vein runs out of new technologies and social conditions alter the demand for their resource, and they die—or become frozen in time as "museums" or tourist attractions.

If, on the other hand, there are reasons why the city in such circumstances cannot be allowed to die (as, for example, in the case of Montevideo [Jacobs 1984]), the cost of keeping them alive is enormous. Typically the cost is paid by people who live in poverty and squalor, and, eventually, the cost of maintaining order is paid for by those who support and equip military and paramilitary organizations who keep the city "intact" by brute force.

All the while the single-resource city is functioning successfully its economic accounts are satisfactorily balanced. In fact it could be thought to be a healthy, growing, thriving place. The entropy debt that is piling up is hidden until the sustaining flux is interrupted. *It* enters the *economic* accounts only when crisis occurs.

The general point here is that the material conditions for the emergence of an economy as a dissipative structure are potentially many dimensional, and no one-track account of the rise and solidification of an economy is to be expected. To put the same point in other terms, an unstable, temporary-looking structure arisen out of the flux of one gradient-tracking trade could be all it takes to ground the stability of a long-lived structure in the flux provided by a newly arisen gradient-tracking trade.

The Boundaries of Equilibrium

Those wedded to free enterprise may take superficial comfort in the view that dissipative structures arise "spontaneously" without centralized manipulative control. But this is shallow. The order-out-of-chaos of the market is most like a simple dissipative structure in the relatively spontaneous appearance of small local markets operating in terms of barter and extended barter. Thus it is clear why these small markets are always invoked as paradigms in present theories of the market. But these small markets are extremely modest entropy producers. And the small entropy debt is paid largely by the market participants themselves, and the beasts of burden (wives, etc.) who truck the produce to market. The rise of complex, fully developed economies is entirely another matter. For example, in contrast, the modern capitalist economy, when looked at in these terms, is a strategy premised on an infinite bankroll.

A main criterion of dissipative structures is their time dependence. This criterion surely must be met. Quite evidently, economic systems meet the criterion; but we have to be careful to see clearly *how* they do so. For much of classical, neoclassical, orthodox economic theory makes use of techniques which play down or obscure the essential time dependence of economic processes. In particular, equilibrium analyses treat the process of economic exchange as if it were reversible. Indeed this is the heart of orthodox price theory and its accounting system. Yet, in another sense—still within orthodox theory—the processes of trading *cannot* be reversible. The path to the bargaining locus cannot be retraversed by rational traders. But in any case it is an illusion to think that orthodox economics is about trading. Nowhere does orthodox economics examine the process of trading. It examines only the logical consequences of a set of assumptions: that the bargaining system contains only rational economic men, defined by the standard process of abstraction; that initial assets are such and such for each of them; and that there are no exogenous constraints on the results of trading. We have to ask seriously how idealizations can engage in causal processes. The answer we usually get is that they do not, but that real beings closely approximating the idealizations do, with their behavior, *ceteris paribus,* conforming to the ideal. I have discussed this claim at length elsewhere (Dyke 1979). Here it is enough to say that the insistence upon the bookkeeping of abstractions sets up insuperable obstacles to our inquiry about the processes underlying bookkeeping. We can recall the discussion in chapter 4.

Two particular obstacles need to be mentioned. First, many of the processes we would like to single out and include in our calculations as we try to understand economies turn out to be unavailable—by accident, as it were—if we confine ourselves to the bookkeeping of market equilibrium. For nothing enters market calculations that has not somehow been internalized in the market, has not found an equilibrium price. The participants in the market have primary control over the items that enter the market and get prices attached to them. Secondarily, government at all levels forces market participants to recognize items not otherwise included in the market by taxing, licensing, etc. But the items taxed, licensed, and otherwise controlled are likewise historical artifacts, items that come to public attention in the course of political life. We find, not surprisingly, that a lot of the things we would like to know about never come to the attention of the market system, let alone become internalized in it.

Second, we have to look at the concept of efficiency associated with the standard equilibrium models. Again, the concept is given in micro theory, then (if ever) imported into macro accounts. In particular, the equilibrium claim to efficiency is that for a given set of resources as inputs, the free operation of the market will yield an equilibrium result that can be shown to be optimally efficient, in the sense that no participant in the market can be made better off by a redistribution without some other participant being made worse off. Now, this conception of efficiency has been the subject of extensive and intensive criticism in its own right, which I will not go into here. Rather, I will point out that whatever concept of efficiency we may eventually want to use in our deliberations about comprehensive economic policy, this is among the last to recommend itself. In the first place, the "efficient" equilibrium is a function of the short-range interests and perceptions of a population of the moment. Since any long-range economic policy would *budget* resources over a long period of time, either the momentary equilibrium is irrelevant, since on the usual assumptions market participants have no reason to aim at long-run efficiency, or we have to suppose that the long-range budget is serendipitously in accord with the short-range budgets of the participants in the market equilibrium. The latter supposition is bizarre.

As an example, we can think of decisions about the use of land. Land use issues usually have the following structure: firms, in the normal course of business, attempt to exploit land for whatever resources it offers. When firms compete for given land, the competition is either between two similar firms for the same resource, e.g. a particular mineral, or between two dissimilar firms for different resources, e.g. a mining company and a hotel chain may both be interested in the same piece of ground.

In the former case, the price of the land may exclusively represent the estimate of its worth based on a single resource. In the latter case, the price of the land will represent the marginal benefits of its use in one way over its use in another way (especially, but not exclusively, when the two uses are flatly incompatible). In both cases, the land will be internalized in the market at a determinate price, but there is no need for that price to make sense in terms of the long-run socio-economic concerns we have. The price simply represents the more or less short-term calculations of one or more firms which have an interest in the land. The use of this price for long-term planning is ludicrous. Yet the price is the only legitimated figure that the equilibrium calculations of the market can yield.

There is no doubt, of course, that equilibrium models offer us more comfort than models of stability far from equilibrium. In the first place, equilibrium models make it seem as if the perpetuation of the status quo were up to us; and if we were just careful enough to behave properly we could be sure that we had our lives in our own hands. Talk of stability far from equilibrium is disconcerting, since it sounds like a situation on the ragged edge of collapse. But as we saw in discussing Malthusian closure, everything that lives may be on the ragged edge with respect to key components of its ecosystem. If this were not so, whatever would happen to the theory of natural selection? Are we somehow exempt?

When we say that an economy is not an equilibrium system, but a system stabilized far from equilibrium by material and energy flux, we recommend a modified accounting system: one that insists on examining the flow requirements for, say, a city, and considers the sources of that flow and the consequences of depending on these sources. Our accounting insists that the internal stability—the viability—of the city is directly linked with the available resources and with the ability of the city to generate some of them itself. And we also insist that many important consequences of this interdependence will not show up in the normal equilibrium calculations unless circumstances change significantly. So if someone can argue that circumstances will always be pretty much the same—the material and energy flow necessary to sustain the city will always be reliably present and suitably efficient structures will continue to exist—then he can argue that the equilibrium calculations are enough, and, for example, land use decisions can be made on the basis of the marginal product of the land under present market conditions.

Or, in short, capitalism and its ideology are and always have been ahistorical. And if we think history—as any serious structural change—can be prevented, we can be comfortable with equilibrium bookkeeping. Evolutionary theory in biology, on the other hand, is a theory which has to be historical. The fossil record and everything else we know about the biota of the earth force this. Equilibrium bookkeeping won't work in evolutionary biology. But, then, this means it won't work in ecology; and if it won't work in ecology, then eventually it won't work in economy either. For the second law of thermodynamics insists that we cannot stop ecological change, or seal ourselves off from the changes, except temporarily and at extremely high cost.

We are used to thinking of the material and energy flow from nature through our economy as one we ourselves initiate (by exploiting natural resources), which we then cause to circulate in such a way that the benefit is distributed among us. Since we seem to see ourselves getting more and more wealthy, individually and collectively, over the course of time, it is hard to think of the flow in NET terms— e.g., as requiring a sink as well as a source. Trash, soot, and sludge seem an annoying and inconvenient byproduct of our lives and activities rather than a necessary feature of them. But without a gradient down which material flow can cascade, no dissipative structure can remain stable. Genuine economic equilibrium looms as the heat death of our civilization.

Naturally I am far from arguing that to maintain a stable existence of the form of life we love we ought to promote the wholesale production of waste. *Far* from it. But our existence as dissipative structures defines a space of possibilities for us, and does so rather tightly. We know from the standard thermodynamic analyses of human life that if we conserve and recycle we can lengthen the course that materials follow as they run through our hands. We know that if we use sunlight (and its immediate and inevitable correlates such as wind) we can select a composition of the material flow that gives us a longer thermodynamic horizon. Of course we do not need the resourses of NET to tell us that. But within the frame of standard economics the pattern of decisions concerning resource utilization is set one way, and within the frame of NET it is set another.

The difference is the following: within standard economics decisions are all framed as cost/benefit decisions. The costs and benefits are regarded as resource allocations, including the allocation of our own time and energy, all within the framework of "equilibrium" efficiency. We *never,* except in the most superficial ways, examine the relations between our patterns of social organization and the rate of material flow needed to sustain them. The optimal efficiency is never satisfactorily connected to the time-dependent thermodynamic conditions that really matter.

Skirting Potential Pitfalls

My analysis of the genesis of cities is, like many other discussions of NET and dissipative structures, obviously incomplete. The incompleteness is closely related to two themes that have figures throughout this book: the distinction between bookkeeping and explanation; and the call for robustness. Entropy bookkeeping is just that, bookkeeping, and by itself never yields robust explanatory results. Critics of NET, both friendly and hostile, have pointed this out in one way or another, time after time. And in my view they are absolutely right. We first have to note that the issues raised with respect to NET can be raised with respect to many other fully legitimated analytical techniques. Most techniques yielding holonomic trajectories through phase spaces have a narrow range of questions they can answer for us. If we recall Garfinkel's contribution, we see that this is fully to be expected. Thus there are all sorts of ballistic trajectories that can be constructed with Newton's laws, i.e., that are "consistent" with these laws and, say, the symmetries associated with them. But to know that something is on such a trajectory can be to know little of interest about it. Suppose we see two objects flying overhead and by good fortune know that neither has a method of self propulsion; we know what each weighs; and we know their velocity vectors at two points on their trajectories. We could then predict, for example, where they will land, when they will land, etc. But if we want to know why they are flying overhead, all the ballistics in the world won't help us, except by leading us back to the flight paths' origins, where totally nonballistic questions could lead us to the explanations we sought.

Similarly, when we begin to look at some chemical concentration distributions within a cell and find the phenomenon of standing waves, NET considerations will provide initial enlightenment on the phenomenon, but further analysis of the chemical kinetics, investigations into cell physiology, and even a probing into evolutionary roots may be necessary to answer the questions that concern us. Indeed, any robust results in any one of these lines of investigation may well require findings from the other lines.

And, finally, when we look at cities as dissipative structures we do not automatically find answers to all the questions we want to ask about cities, their genesis, their internal dynamics, and the dynamics of their interaction with the surrounding world. Far from it. All we find from our initial NET account of cities is the shape of the possibility space within which cities grow and survive. Indeed, the constraints and enablements we find for the trajectory of cities are very general, and we have to look to much more specific material conditions for this or that particular city to begin to get a fully useful explanatory picture.

What techniques such as NET do for us is to help us locate regularities, generalities, and contingencies. They guide us to where explanations are likely to lie. The positivist social sciences have always urged the discovery of "social laws." As we have seen several times, if explanation were derivation, then we would indeed be stuck with the bipartite task of finding atomic facts and the laws that bind them together. But as we have also seen, the actual activity of science is much more plural, complex, and sophisticated than the positivist parody of it will allow. The fate of "laws" in the social sciences has not been a very happy one. In a realm of multi-level interactive systems, this is not really a surprise. In such systems, finding a regularity is not tantamount to finding a law. A discovered regularity is merely an indication that further investigation has a chance of unearthing something important.

Furthermore, we have to remember that in dealing with complex systems the most useful form of a "law" is very likely to be its extremal form. So, in the main, laws will not be associated primarily with positive regularities, but with systematic exclusions. Stones do not fall up, metabolic pathways do not achieve 100% efficiencies, and, perhaps, cities do not thrive if they fail to develop new work which multiplies and diversifies the division of labor within them (Jacobs 1970, ch. 2). Unraveling the grounds of these systematic exclusions, assessing the imperatives they occasion, and articulating the processes taking place under their constraint are all complicated, arduous tasks. It seems ironic to me that the social sciences, so desperate in their attempt to emulate the natural sciences, opted for the illusory positivist surface and failed to learn the real lesson. This lesson is that science is hard work. The social sciences should look at what it really takes to establish scientific results, or at what goes into understanding something like cell metabolism. For, since social systems are at least as complex as the cell, it is very unlikely that one or two bright ideas, one analytic technique mechanically applied, will yield an understanding of a social system.

Of course this is not at all to condemn *all* social science out of hand. Among other things, that attitude would deprive me of access to material I have depended on from time to time in this work. But it does give us a new stance from which to reevaluate much of what the social sciences have offered us so far. And if we find fault with the social sciences, we have to be sure we know where it lies. For example, it has to be pointed out that the social sciences have never enjoyed the same degree of autonomy as the natural sciences. Perhaps they never can, since the phenomena they investigate have such immediate relevance to policy-making. Furthermore, we have to remember that the history of investigative social scientific practice has its roots in normative social theory—or, better, in an investigative tradition in which the radical distinction between normative and positive theory is not made. Indeed, it may never be possible to enforce this distinction stably. Barry Schwartz maintains, for example, that the social sciences always have been and always will be the field upon which the battle for a conception of human nature takes place, and he may well be right (Schwartz 1986). If so, the complexity of the task of understanding social systems is extreme, requiring constant attention to ideological frameworks that may never be completely eradicable, but which need to be treated as we would treat other ineradicable boundary assumptions. Under these conditions it is a treat to find work which approaches the degree of solidity we are used to finding in the natural sciences. The next chapter will begin by looking at such a work, giving a bit of concrete content to some of the preceding generalities.

CHAPTER 10

Toward a Social Dynamics

The success of the biological investigation that followed upon the Darwinian beginnings depended, as all fruitful investigative programs do, on the ability of successful answers to generate new, well-formulated questions. I have been stressing throughout this work that new, well-formulated questions have been generated within the Darwinian tradition, forcing investigators within that tradition to expand and open up their explanatory repertoire in the face of complexity which outruns the exclusive reliance on a totalized adaptationism. The expansion of explanatory resources has been exploited by researchers at the bench and in the field who have had to complexify their investigative methods to keep pace with the complexity of the questions they have learned how to ask. Even as they move beyond Darwin, our debt to him is well paid, for his importance as the founder of the research tradition continues to be honored, and we can find a solid place for natural selection among complex determinations of differential stabilities of systems relative to their surroundings.

The current challenge, the one that motivated me to write this book, and that motivates the sociobiologists, is to take what has been learned from the investigation of the evolution of organisms and see if we can exploit it in investigating the evolution of social systems. In the last few chapters I have tried to take steps in that direction. However, it must be noticed that the questions (well formulated, I hope) that guided the discussions of the last four chapters arise not from the dominant traditions in social science, but as extensions of research in evolutionary biology. So as we move on we must recognize that the question of well-formulated questions has to be addressed immediately and continually. We must constantly ask whether we are asking the right questions. The sociobiologists are quite aware of this but sometimes think of themselves as missionaries to the social sciences, teaching total reformulation of agenda.[1] My attitude is somewhat different. Just as I want to insist on preserving the best of the findings of the neo-Darwinians, despite my belief that we must go beyond the limits of their "paradigm," so I think that we have to respect solid findings of the social sciences despite the origins of their research traditions being questions that not only define and constrain their activities, but, from some points of view, may even be the *wrong* questions, in that they defeat rather than further our understanding of social systems. The orthodox social sciences, in other words, must be retained within the dialogue even as we advocate new sorts of investigations.

Consequently, as promised, we will look at what I think is an exemplary work in very orthodox economics (Kelley and Williamson 1984), even as I suggest some of its limitations. After that we will look at possible expansions of the investigative field suggested by the lines of thought we have developed.

What Drives Third World City Growth?

Not everyone keeps up with the theoretical literature on Third World economy, geography, and demography, but nearly anyone with an ounce of interest in the overall development of our planet is aware of the (apparently) enormous growth of the cities of the Third World. Indeed, if nothing else has come to our attention, the prediction that within a few decades Mexico City will be the largest city in the world has surely surfaced in newspapers and on television. This should interest everyone from geologists to used car salesmen. While Mexico City may be an extreme example, the issue of city growth is of central concern throughout the Third World. This is sufficient to explain the wealth of theoretical work being done to understand this growth, with an eye, of course, to controlling it or managing its consequences. Perhaps the most rigorous attempt to understand the growth of cities in the Third World is that of Kelley and Williamson.

I want to show how serious investigations in this area generate heuristic issues just like those we encountered in our consideration of biological evolution. In particular, I focus on issues of systems boundaries and the possibility of utilizing techniques arising from the thermodynamics of open systems of the sort explored in the last chapter.

The authors identify their theory as "A Dynamic General Equilibrium Approach."[2] Their strategy is to build a model of the Third World economies which carefully articulates parameters, endogenous and exogenous variables, production and consumption functions, price relations, and dynamic equations. It delineates sectors of the economy in the way that is most relevant to the circumstances they propose to study (e.g. skilled and unskilled segments of the work force both in the cities and the agricultural regions).

Since it is an *economic* model, certain boundary decisions and choices of abstraction are naturally made prior to the model building itself. As usual, these decisions and choices concern motivational assumptions and assumptions about internal equilibria. The authors are very clear that such assumptions constitute decisions about explanatory closure, something I emphasized in my discussion of game theoretical explanation. Thus they say:

To begin with, the multisector general equilibrium model developed in the pages that follow possesses a high degree of what the specialist calls "closure." Most input and output prices are determined endogenously. Neoclassical production functions are assumed and price-responsive aggregate demand functions are implied by the household demand system postulated. A period-by-period equilibrium is imposed on the economy whereby factors move between and within sectors, minimizing rate of return and earnings differentials, subject to various institutional constraints and imperfect information about the future. Optimization at the microeconomic level is imposed on firms and households that independently maximize returns and utilities subject to budget restraints. (p. 13)

This set of closures is virtually canonical for general equilibrium models, deriving from a long tradition within orthodox neoclassical economics (the same tradition, in fact, that gave rise to the theory of games). As usual, certain other closure assumptions are required as the model is applied. We will not consider them, since it would require a much longer exposition than is possible here. Some of the consequences of the main closure conditions will be explored below. Here we can note that imposing closure conditions sometimes has some pretty dire effects. The following case will be of some interest to evolutionary biologists. In a working paper prepared for the World Bank (Mahar et al. 1985) a group of authors addressed themselves to questions of population growth and human carrying capacity, which was defined in terms of natural food supply. Now, the necessity for closure is particularly acute here, for in the age of global economies the habitat of any human being is, in a real sense, the whole world. The closure conditions chosen are, first:

In sum, with no changes in technology, capital, or location, the community could vary only its numbers and its leisure/work time allocation. Under these highly limiting conditions, the carrying capacity of the land (in combination with the work/leisure preference function) completely determines the community's size. (p. 10)

The additional, and related, assumptions are that boundaries remain unchanged and impermeable, that is, "nation" is fixed, and there is no international trade. In short, carrying capacity can be defined so as to determine population changes only for closed systems. But since there are no longer any such closed systems, we must say either that the concept is useless, or that it is useful, but only in hypothetical circumstances (which never hold). Biologists can here assess the roots of their relief when they discover that they are dealing with a system of "island biology." We can hope that the closure assumptions of the Kelley and Williamson model are not as debilitating as those of Mahar et al.

Next to be identified for the Kelley and Williamson model is the set of well-formulated questions that are to be answered in terms of the model. The paramount question is the obvious one: "Why are the Third World cities growing (apparently) so fast?" Obvious subquestions concern whether cities are growing because of increased birth rates or because of positive net immigration (or some combination of both). The answer seems to be that the cities grow because of net positive immigration, so we then naturally want to know how this immigration pattern is related as a dependent variable to other variables identified by the model as independent variables. Finally, there is a "metaquestion" raised and answered in the affirmative by the authors: "Can the complex social process that produces urbanization and city growth be modeled?" (p. 178) This requires close examination.

The model is sufficiently rigorous to support some counterfactual comparisons, which Kelley and Williamson rightly point to as a major advantage. That is, they are able to ask reasonably, "What could we expect to be the behavior of city growth patterns under conditions other than those that actually obtain?" This is particularly important for their purposes, since the formation of OPEC during the period that they examine introduced a serious discontinuity in fuel prices and availability, which has to be "factored in" very carefully. In addition, the question of the "normality" or "abnormality" of the growth rate of Third World cities requires an answer based on precise comparisons of a counterfactual nature.

The Results

In something of a contrast to hypotheses advanced by others, Kelley and Williamson conclude that the growth of Third World cities has as its major independent variables unequal growth in rates of productivity between various sectors of the Third World economies. That is to say, the introduction of industrial capitalism, either in its classic or state-socialist forms, locates the more productive—and increasingly productive—means of livelihood in nonagricultural pursuits, and people gravitate (by the closure assumption) to these pursuits, which (by the closure assumption) offer higher wages. (A close reading of the model shows that the major sector dichotomy is not urban/rural, but agricultural/nonagricultural.) Other theorists had suggested rural land scarcity, sheer population growth, or the influx of foreign capital as causes of urban growth. Only the last of these figures (indirectly) as a cause in Kelley and Williamson's account.

Another way to put the main results of the model is to say that urbanization is a natural result of industrialization. As simple as this formulation seems, it is quite sound and is reinforced by the authors' demonstration that the rate of urban growth in the current Third World is quite comparable to the rate of urban growth in England during its most intense period of industrialization. In other words, the Third World cities are not growing at an abnormal rate if normality is relativized to the appropriate stage of development. Furthermore, on this basis, alarmists who worry about the present growth rates of Third World cities continuing can regain their calm.

> By unbalanced total-factor productivity advance we simply mean that technological change is usually more rapid in the modern, urban-based primary products sectors. Traditional service sectors, of course, also tend to lag behind. The size of the bias and the magnitude of the unbalancedness varies across nations, but it has been a technological fact of life since Britain's first Industrial Revolution, and in spite of past agricultural revolutions and contemporary Green Revolutions. Chapter 5 argues that the unbalanced rate of technological

progress in the Third World was the key favorable condition that accounted for the unusually rapid rates of city growth in the 1960's and 1970's. It follows that if the productivity slowdown currently characterizing the industrialized nations spills over into the industrializing Third World over the next two decades, Third World city growth rates will retard as well. (p. 183)

Others might be inclined to express the point in terms of the changing composition of capital, and fail to find solace in the thought that the Third World is destined to go the way of Britain.[3]

Search for Perspective

I see no reason to quarrel with the main conclusions of Kelley and Williamson, though I am well aware that within their own field there is still serious disagreement about their findings. I am more concerned about the light their analysis sheds on the heuristic strategies that dealing with systems as complex as Third World economies requires. So, I will assume that their model does indeed provide good results and proceed to integrate it with my analysis.

The obvious first question is how the Kelley and Williamson model fits with the conception of cities as dissipative structures. After all, theirs is a general equilibrium analysis, and dissipative structures are defined as far from equilibrium systems. Yet there are many NET conclusions to be drawn from their findings, just as there are serious questions to pose for them from a NET point of view. First of all, on the basis of their analysis the growth of cities appears to be the response to changes in flux conditions. For example:

Table 5.4 has already confirmed that imported fuel and raw material abundance helps explain the rapid city growth in the Third World up to the OPEC watershed and the slowdown thereafter. Similarly, Table 5.4 also confirmed that world market conditions for manufactures were also highly favorable forces contributing to high rates of city growth in the pre-OPEC period, the reversal of those conditions accounting for much of the slowdown thereafter. Indeed, Table 5.4 suggests that world market conditions for manufactures were a more important ingredient of rapid city growth up to 1973 than was the relative abundance of fuel and raw materials. (p. 138)

This is just what would be expected of a system maintaining its stability far from equilibrium under variable flux conditions. Under conditions of increased flow-through (occasioned *both* by gross availability *and* the gradient-creating effects of manufacturing conditions, which result in a sink being made available) the city grows. It also complexifies. This is the conclusion we can draw from the fact that movement to the city is accompanied by shifts in demand from unskilled to skilled labor and increased investment in the training to provide that skilled labor. From this perspective Kelley and Williamson's emphasis on the analogy between current Third World economies and the British economy during its intense industrial development becomes even more fruitful than their use of it argues. It becomes an invitation to go back to economic history (again Braudel is a benchmark) and see if the developmental patterns exhibited by Kelley and Williamson hold more generally.

Furthermore, we find another characteristic of dissipative structures, a differential sensitivity to exogenous perturbations depending on the system's location on a growth and development trajectory.

> In summary, initial conditions *do* matter. Urban performance of countries in intermediate stages of development appear to be more sensitive to exogenous shocks than are those at very low or very high levels of development. This statement holds, of course, only for *urban* attributes. (p. 115, original emphasis)

On NET grounds we could hypothesize that countries at low levels of development have not yet completed the transition to a new productive and integrative regime—a new division of labor—and are still relatively stabilized under old conditions. Similarly, highly developed countries have been integrated into the new regime and have developed sufficient internal organization to buffer flux variations. The same phenomena are, after all, well known from studies of local ecosystems undergoing change under the impact of climatic or other changes. However, this analogy, and reference to changes in the division of labor (and the composition of capital) ought to alert us to the necessity to go beyond what the Kelley and Williamson model can provide. For example, while the urban growth patterns of industrializing Britain and the industrializing Third World are (now expectably) analogous, each area has its own particular history. The general similarity is accompanied by many differences resulting from the differences in the underlying structured structuring structures that define the possibility spaces within which development takes place.

The general equilibrium model is constructed in such a way that it inevitably obscures these differences. Differences in developmental conditions are assumed and abstracted away by the very closure conditions that make the model possible and useful. The model can explain the shape of the trajectory of, say, the city economies of the Third World, but because of its generality cannot determine the particular ways in which the growth and complexification of the cities will proceed. This is not a criticism of the model, but a limitation that becomes a problem only when the forces abstracted away by the model are those that form material conditions strong enough to perturb the future trajectory outside the limits the model allows. The assumptions of the model presuppose that the structure of the world economy will remain, within reasonable limits, roughly as it is now, thus that the international political situation will remain roughly within the same sorts of limits. It also assumes, though this is less straightforward, that the Third World economies will continue to develop so as to accommodate themselves to the conditions, style, and pattern of the dominant economic world of the developed countries.

This last point leads us to consider again the profound social transformations hidden in the analytical apparatus of Kelley and Williamson's eight-sector model. Urbanization and attendant changes in the division of labor and social organization that the model analyzes constitute a revolutionary transformation of the human ecology of the developing nations. The closure assumptions of the model speak of the optimizing motivation of the "household" when the "household" has no concrete content. Remember that the developing country is, as it were, operating in the neighborhood of a bifurcation point. The stability requirements of the old way of life may be very different from those of the new one being implanted by the international economy. The "household" stable as a substructure in the old way of life may fail to be stable in the new regime. Despite people's attempts to buffer their move to the city by reestablishing extended family enclaves in the low-rent housing available to them, or built out of the junk available to them (Hall, 1984), the household's internal structure—breadwinning, child care, and cohort relationship—may change radically. (Were grandma and grandpa left back on the farm? If so, what role did they perform in the rural setting? Was it essential to the stable household? Who performs that role in the city setting?) In short, in the trajectory toward industrial organization, in the building of the structures necessary for the new way of life, old social structures may well become destabilized. That too is to be expected, for the ecological economy available as material for the new way of life is not an economy of abundance. Work is shifted from maintaining and reproducing the old social formations to developing new ones.

This bears on the model's closure assumptions directly. In traditional ways of life in the Third World it is far from clear that household decision-making units produce rational optimizing decisions of the sort required by the closure condition *in the economy modeled by Kelley and Williamson*. On the other hand, if their model is explanatory, as it certainly seems to be, this means that households *reconstituted in accordance with the new set of necessities occasioned by economic development* are indeed making decisions within the bounds of closure. We thus return to an earlier question: "Can the complex social process that produces urbanization and city growth be modeled?" The answer we must now give qualifies the flat affirmative initially given by Kelley and Williamson. They have modeled one of the important surfaces of Third World city growth. They have shown the factors most responsible for city growth once the transition to the accounting system of the new way of life has been instituted, once the actors in the developmental drama have been disciplined to the new regime. The model seems quite robust in such circumstances, and a clear advance over the partially theoretized educated guesses that came before. But the model is *not* robust as an account of the entire "complex social process."

The reconstitution of "households," not as abstract entities but as real social substructures, is left out of the accounts. The costs of the transformations required are never directly internalized in the economy defined by the model, despite the fact that this model goes far beyond most of its competitors in capturing the true distribution of costs and benefits of development. But we know, on elementary thermodynamic grounds, that the reorganization of the society in the service of the new economy proceeds at the expense of increased entropy somewhere in the surroundings. Some of the "entropy debt" is paid off in terms of straightforward energy degradation, but not all. Some of the expense of the new organization is paid off in the dissolution of old stable societal infrastructures unsuitable for the new way of life. Now, we can lament this, or chalk it up to progress. That is not at issue here. The point is that in order to model "the complex social process that produces urbanization and city growth" these other considerations must be taken into account. Investigative resources beyond what Kelley and Williamson provide must be utilized. Failure to do so will eventually create unfortunate illusions.

In addition, the model is highly constrained by the requirement that a quasi-static movement be established through periodic equilibrations between the various sectors. This is absolutely essential for carrying out the computations. However we should not lose sight of the fact that the system as a whole is far from equilibrium, and its dynamics (including the migrations of the labor force) depend on maintaining gradients. Biologists will immediately recognize in this remark an analogy both to the dynamics of ecosystems and the internal dynamics of both cells and organisms. The investigation of, say, the metabolic dynamics of a cell surely requires the identification of a host of internal equilibria (such as the various equilibria between enzymes and substrates) which are in a continual process of formation, disappearance, and reformation in response to internal needs, external flux conditions, and various signals from other cells and other parts of the organism. If such equilibria were not present and discoverable our understanding of the workings of the cell would be greatly impaired. Yet, again, we know that the cell is an open system far from equilibrium. Consequently we must reassess the significance of the play of internal equilibria. Just so in the economic system in question here. The analysis of the formation and reformation of equilibria is indeed one of the important parts of the investigative task, but it ultimately only has significance in the context of the economy as an open system which as a whole *cannot* be in equilibrium either in all internal respects or with its environment. As opposed to some other macroeconomic models, I think that general equilibrium analysis has the capacity to be integrated into a more general theory of open systems. That remains to be seen.

There are two remaining tasks: to explain my reference to the potential creation of illusions, and to summarize some general consequences of extending evolutionary biology to the analysis of social systems.

The Costs of Discipline

The discussion of the transformation of old ways of life by the development of the Third World raises the question of the conditions that are necessary for an economic system to establish and maintain its stability. Underlying material social conditions determine part of the system's possibility space. One obstacle to examining these issues is the *a priori* ideological assumption that everyone is quite naturally a rational economic man, perhaps unable to realize this natural bent because of distorting social arrangements. So when the opportunity to become a fully realized rational economic man arises (as in the industrial development of the Third World) everyone will immediately adopt the required behavior without transition cost and become a well-oiled cog in the gleaming new economic machine. But, as we saw earlier, this assumption is extremely problematic. In general, disciplining the members of a society into their required economic roles will have costs. The illusions I have referred to are illusions about the costs

of the internal order required for the stability of a social system primarily organized in terms of, say, the microeconomic closure conditions assumed by Kelley
and Williamson.

When we turn to an examination of the conditions of internal order we also
have a place to start, it seems to me. The project begun by Michel Foucault (especially 1965, 1977, and 1981) promises to be extremely fruitful. His analyses of the
evolution of order and power in the modern world fit extremely well with talk of
possibility spaces, structured structuring structures, constraints and closure. The
particular point we want to retrieve from Foucault is how political power impinging on both individuals and on economies is to be conceptualized. For, lurking at
the boundaries of orthodox economic analyses such as the one we have been looking at is an inherited conceptualization of power and sovereignty which is fundamentally monarchical.

The implications of this for economic models are, first, that government—
especially in its "interventionist" activities—is thought of as an essentially autonomous "cosystem" imposing exogenous constraints on the economy, some of
which will be internalized by the economy and others of which will be treated as
externalities. The tradition of treating government this way in economics is very
old, stemming from the beginnings of classical economics in a polemic against
mercantilism. I will leave this aspect aside here except to remark that one of the
differences between general equilibrium models and, say, Keynesian models is the
way they handle the boundary issues between the political and the economic subsystem. Since we are concerned at the moment with the issue of internal order, I
want to focus on a second aspect of the conceptualization of power as monarchy.

As Foucault shows in great detail, when European societies overthrew mon-
archs as authoritative law-giving sovereigns, they simply redistributed the sovereignty of authoritive law-giving to others. The body of the king was reconstituted
in the body politic. In our own time the conventional wisdom about the Western
democracies is that the redistribution of sovereignty is complete, and that all adult
citizens possess it. They are absolute in their own private domains. They are
minimonarchs. Their homes are their castles. The dynamic of political power is
conceptualized in terms of negotiation and the striking of bargains between minimonarchs. No other account of social dynamics is consistent with the ideology
of democracy—requiring each person to be sovereign, that is, a minimonarch.

Foucault attempted to show how this monarchical model fails to account for
the dynamics of power of various key institutions within modern Western society
and ultimately fails to account for the deployment and use of power *anywhere* in
present-day life. At this point, the main strand of discussion continues in his
work, not mine; for a detailed layout of the development of the dynamics of
power in modern society we should turn to Foucault's writings. Here I will only
lay out some strategic considerations to keep in mind as we think of the social
transformations required within the developing nations. The microeconomic closure conditions applied by Kelley and Williamson require that individual households behave as minimonarchies within the economy, and, indeed, for stability
to be maintained in such a system such behavior is quite probably required.

If such a model explains the behavior of people in the Third World as, for example, a migration along wage gradients, then the initial inclination is to conclude that they are operating as sovereign maximizers. But there is not sufficient closure for such a conclusion. The situation here is exactly like that discussed in chapter 4. The bookkeeping of sovereign maximizers is indistinguishable from the bookkeeping of highly constrained agents with no alternative but to behave as sovereign optimizers would behave. Without a further analysis of the possibility spaces of the people in question, and a full analysis of the determinants of the shape of that space, there are many things we will not know about the very success of the model under its conditions of motivational closure.

The natural response at this point is to search for sources of power of a monarchical nature which may be impinging upon individuals, and then, if we find none, to conclude in favor of individual sovereignty. But this is precisely the point at which Foucault can help. For he points to what he calls "the microphysics of power." For our purposes we can understand this as the web of intra and inter-systemic interactions within a levels-interactive modular array, where the interconnections of activity and response are not, typically, instances of gross one-dimensional control. It is hardly my job here to articulate the microphysics of power within the new ways of life emerging in the developing Third World, but the current state of theorizing about such matters makes it worth pointing out that an examination of such matters must be placed on the agenda. In particular, such an examination is required if we are to fill out the analysis provided by Kelley and Williamson. Discerning biologists will see in this shift from monarchical conceptualization of power to a microphysics of power an analogy to the meteoric rise and subsequent fall of the Master Molecule conceptualization of replication and, particularly, transcription. The advances of the last few years in that area have largely been the result of a fuller appreciation of the complex interactive dynamics of these processes. The call for a consideration of the microphysics of social power is nothing more than an invitation for social theorists to learn from the recent history of molecular biology.

We needn't leave everything in abeyance, however, for we can also link up with the considerations of chapters 6 and 7. It seems to follow from those chapters that anyone interested in maintaining a particular order would be wise to arrange for the social information space to be as small as is consistent with the performance of the activities necessary to produce and reproduce the system. To pursue the metaphor of a few chapters ago, to perpetuate a game with the least amount of effort, get the players totally involved in it and therefore not thinking, for example, of what they would rather be doing. The optimally efficient way to do this would be to make playing the game the only thing the players knew how to do, and to keep them happy doing it.

Two classic strategies for making people resign themselves to the game they are playing are, first, to convince them that the game is written into the order of nature and cannot be changed; and second, to convince them that the game is the culmination of a process of perfection beginning with the dawn of human history. Most cultures have employed these strategies willy-nilly. In our own culture both strategies are represented best by the claim, first enunciated in an appropriate mix

by the Physiocrats and Adam Smith, that economies consisting of free independent entrepreneurs organized in a market was both natural and best. The same
line of thought seems to persist in the more extreme versions of current sociobiology. Right or wrong, aware of its historical role or oblivious to it, any effort
to link biology and society—including both the one I am engaged in and orthodox
sociobiology—has the potential to be important ideologically as it contributes to
the self-conceptions entwined in the microphysics of power. I would sooner face
that fact than ignore it.

Wrap-up

The discussion of Kelley and Williamson invoked themes that have recurred
throughout this book: exclusive reliance on equilibrium models, one-dimensionality, synchronic reduction, natural laws, skepticism-based epistemology, linearity, and the overriding value of simplicity. We saw that Kelley and Williamson
are highly sensitivity to such issues as explanatory closure, multiple access,
robustness, and potential nonlinearity. Just as in evolutionary biology, here good
investigation threatens to outstrip the assumptions under which the investigations were conducted. This is again a central point since it again points up the
way in which the history of investigative practice is itself nonlinear. The progress
of investigative practice is not linearly accumulative. Advances within an investigative practice characterized in terms of structures, ontological commitments,
and scruples of one sort can themselves lead us, reasonably, to modify the very
structures of investigation that led us to the advances. Thus the reductionist, atomist canons which undergirded the program that led to what we now know in
molecular genetics have led us to the point of relaxing or abandoning these very
canons. This is demonstrable not on the basis of a detached *a priori* critique of
biology and chemistry, but on the basis of the newer, more sophisticated practices
of biologists and chemists themselves as they apply themselves to new tasks at
the cutting edge of their fields and try to integrate their results. What they learned
on the basis of the old assumptions remains solid for deployment in the light of
new structured structuring structures of investigative design. The shape of explanatory space changes—to reinvoke Garfinkel—and the most important consequence is that the *next question* asked as you advance along an investigative pathway is different than it would otherwise have been. If progress is to take place, the
old successes must be understood in a new way. As we have seen several times,
we can still admire the generative capacities of old presuppositions even as we put
them behind us.

   Along these lines, it is crucial that talk of developmental constraints, stabili-
zation of information, and, at the boundary, natural selection, be extended—with
care and critical caution—to the social realm. This is the integration of thought
about the dynamics of complex systems I have been trying to achieve all along.
But we can see some important differences between the biological and the social
systems. In particular, if we want to take dissipative structures as a basic starting
point, we have to note that the stability conditions for social systems can be very

perplexing. Within levels-interactive modular arrays it may happen that the stability conditions for some of the modules are incompatible with the stability conditions for other modules. This is likely to be true especially in times of transition, such as now in the Third World, where old ways of life are struggling to retain a foothold in the presence of new ones. Systems of marriage and kinship, local organization, and the like are all threatened with extinction. As we saw, these facts delineate the limits of our ability to make certain equilibrium assumptions. If, as Kelley and Williamson tell us, the forces of change involve the flux associated with the international economy, then we have a situation in which internal organization is, momentarily at least, incompatible with flux conditions.

Of course in the biological world flux requirements and internal order can be incompatible too. When they are, "evolutionary events" take place—evolution or extinction. But when incompatibilities arise in biological systems they give rise to fairly evident approaches to systems boundaries—Malthusian closure. In this way the adaptive capacities of organisms or particular individuals within a population can be tested. Explanatory closure can be achieved, at least from time to time. But explanatory closure is much harder to achieve in social systems. Take, for example, the discussion of game theory from chapter 4. Game theory was, of course, invented to explain human behavior. From the discussion in that chapter we have a good sense of what it would take to generate explanatory closure for a game-theoretical explanation of some human action. It would require that life be reduced to a one-dimensional game. Game theory assumes precisely this closure for its success. Now, the interesting thing is that since human possibility space is partially humanly determined, we could try to *create* game theoretical closure conditions writ large, just as we create them writ small when we play bridge. Now that would be no small task. As you may remember, the fact that there are one-many, many-one relationships between bids and hands in bridge means that we can only establish loose closures *even within that tiny confined space*. But if we could create humans who were constantly under closure at the margins of their financial and other budgets, we could achieve game-theoretical explanatory closure of a sort. That is, if we could make them maximizers at budgetary margins in all phases of their life, we could totalize the system of game-theoretical explanations. This, of course, is what Kelley and Williamson have done at the theoretical level, and what modern consumer society tries to do at the practical level.

So, to generalize the discussion somewhat, suppose totalization has been achieved in terms of this closure. The result will be a tightly disciplined one-dimensionally ordered systematization of human life. It will constitute a very accessible system of internal order whose flux conditions can then be assessed. In fact, the history of the American economy since World War II can probably be seen as an approximation of a systematic exploration of the flux requirements for such a system. (It is only an approximation because totalization in this one dimension has not been achieved.)

During this period (and, in fact, throughout its history) the American economy has been subjected to fluctuations in the background flow of sustaining material and energy both quantitatively and qualitatively. Knowing this, an energeticist would expect the system to slip into a regime of limit cycles as internal changes lagged behind the ambient fluctuations. Is this what economic and business cycles are? If it is possible that they are, then we have all the more reason to examine the flow requirements we are imposing on our lives by trying to totalize it in one-dimensional terms.

But to do this we would have to stop thinking of the boundaries of economic life as if they could be decoupled from the ecosystematic context that provides the immediate material conditions of background flux. We would have to regard the sensible management of resources and the prudent management of waste not as Protestant frugality, but as a basic constraint on human life. This is what it means to establish a continuity between biology and society—not the reduction of social life to biology, but the recognition of the shape of the possibility space through which social life has to pass, a space with ecological determinants.

Some facets of social order will be rather easy to quantify in terms of reasonable energetics parameters. The analogy with keeping the energetics books on metabolism in terms of ATP is relevant here. Others will be much more difficult to quantify (Adams 1975). In particular, the subtleties of habitual and stylized human interaction are difficult to put into energetics terms. This is one of the reasons information theoretical conceptions are particularly intriguing candidates for exploitation here. In the clearest instance, the generation and transmission of *technē,* the discursive practice of technology, constitutes an expansion of the information space available to a people. This is well exhibited in Kelley and Williamson's analysis. The expansion changes the efficiency of the utilization of available energy and material resources. So it is obvious that this increased efficiency will show up in the total energy/entropy accounts of the system. These accounts will then give us a secondary measure of the information content of the society. The same may be true of an entropy measure of social cohesion. That is, it is reasonable to expect a group of people who get along with one another to be more efficient in maintaining order than a group divided by dissension. So gross measures of cohesion may be available. Yet it would be very difficult to quantify the relevant differences in any more refined way. Given the general difficulties in quantification, we may have to be satisfied with ordinal measures of cohesion relativized to a particular set of energetic circumstances. In this we would be no worse off than the orthodox microeconomists who must settle for an ordinal base for their calculations (Dyke 1979).

Furthermore, from the "internal" point of view of members of the social system, many of the phenomena we would like to quantify may be considered qualitative rather than quantitative. This could often have the consequence of generating disputes about the very concept of efficiency employed in the first place. This is precisely the situation, for example, in any dispute pitting a nonequilibrium thermodynamic conception of efficiency against a Paretian conception. While we may know that the system is, thermodynamically, not an equilibrium system, it may be part of the system of discipline and order *within* the system for people to think of it in equilibrium terms. They may, for example, think of economic equilibria as devices by which comparative well-being can be assessed and managed—and act on the basis of that conception. This is certainly true of most

highly developed economies at the present time, even those that think of themselves as socialist.

But now, how do we discern the size and shape of the possibility space through
which a socioeconomic system has to pass? How do we determine its stability
conditions? How do we make educated judgments about the dynamics of its
future evolution? The answer to these questions is an interesting one from my
point of view. For, just as I claimed that a serious modification of the old Darwinian program has emerged from within that program itself, so I think that the
accumulated body of knowledge in economics and economic history can go a long
way toward answering these questions—*if* they are posed. To use a thought from
Garfinkel, again, we have accumulated a lot of knowledge in search of answers to
the wrong questions. Yet there are some people asking the right questions nowadays, and formulating answers very much in line with the explanatory grid I am
advocating. For example, two works of equivalent scope appeared within a few
years of each other, Carlo Cipolla's *Before the Industrial Revolution* and Braudel's
Civilization & Capitalism 15th–18th Century, both by excellent historians. The
difference between their works is not just that Braudel's is about five times longer.
The difference is the extent and shape of the explanatory resources deployed in
the two books, and, correspondingly, the questions they ask and attempt to
answer. The "facts," so to speak, are there. Generations of diligent archivists have
rooted them out for us. Braudel's explanatory grasp of them is superior. A good
part of its superiority lies in the fact that it opens out a possibility space for further
research. To a large extent this is due to the research agenda of the *Annales* school,
explicitly opening out historical research to include findings from as many disciplines as possible (Stoianovich 1976). The result of this incorporation of many
investigative fields is that Braudel achieves an impressive degree of the sort of
multiple access I have been advocating all along.

Braudel's project, and its remarkable degree of success, is an especially important exemplar in the context of the heuristic recommendations of this book
because he takes on the task of examining the long-term evolution of a large complex system of systems, the world economy prior to the Industrial Revolution.
This makes his work more closely analogous to that of the evolutionary biologist
than is usual in the human sciences. Furthermore, his sense of the long-run
dynamics of world economies can easily be fit into the apparatus I have set out,
the apparatus of structured structuring structures, interactive levels, and even, I
would say, the determining constraints on systems far from equilibrium. For
example, Stoianovich points (more or less accurately) to:

> ... Braudel's conception of three main arrangements of duration: duration at a
> quasi-immobile level of structures and traditions, with the ponderous action of
> the cosmos, geography, biology, collective psychology, and sociology; a level of
> middle-range duration of conjunctures or periodic cycles of varying length but
> rarely exceeding several generations; a level of short duration of events, at which
> almost every action is boom, bang, flash, gnash, news, and noise, but often exerts
> only a temporary impact (1976, p. 109).

In a longer treatment (forthcoming) I would, of course translate Stoianovich's
poetic excess into the terms used throughout this book, and show with as much

exactness as possible that what we learn from successful attempts to understand biological evolution can provide an extremely powerful integrative historical research heuristic.

We can end by addressing a question made inevitable by the history of social thought. Does this proposed integration of biology and social theory turn out to be a rehash of the old organicist theory of society? No, although there are some similarities. All the organicist theories I have ever studied have depended on some concept of teleology to hold them together. Often it is couched in a metaphor of a conative search for equilibrium. The model presented here has no such dependence. The metaphor of the possibility space says nothing to guarantee smoothly functioning social organisms adapting to historical circumstance. It also says nothing about romantically conceived life-cycles for civilizations. This is hardly the century for utopianism. One of the dimensions of *our* possibility space is the tolerance of human bodies for ionizing radiation; and we do not even know at the present time if we can keep *that* part of our future open. So equilibrium and teleological assumptions are absent. They were a conspicuous feature of all early organicisms, and if they are required for *any* organicism, then what I have been presenting is not organicism. Of course in those terms I do not even hold an organicist theory of organisms.

On the other hand, I do hold that socioeconomic systems evolve, as organisms do. I think that there are systemic interdependencies within socioeconomic systems, as there are within organisms. And I think that the research strategies developed in biology for the analysis of complex evolving systems are very fruitful for the analysis of social systems. But a determination of how far these analogies go depends not on the *a priori* assumption that the analogies must hold to the letter as far as we could ever push them, but on carrying out the sorts of investigations I have been urging. Let the games begin.

Notes

Chapter 1

1. Similar thoughts are well expressed by Robert Rosen in his Ashby Memorial Lecture, "The Physics of Complexity," which appears in Trappl (1986).

2. The emphasis on *radical* mismatches is the distinctly Malthusian contribution. An entirely different picture emerges if conditions of moderate mismatch are contemplated. For then we would have a Humean, rather than a Malthusian situation to examine and might expect internal adjustments of population size, habitat-sharing, etc., to be at least plausible candidates for dynamic responses within an ecosystem.

3. This is not to say that all monarchical or paternalistic elements were successfully eliminated, as we will see in later chapters. For the pervasive strength of these elements see Keller (1983 and 1985).

4. . . . as it became obvious in *McLean vs Arkansas Board of Education*. A challenging, perhaps infuriating, account of the case is contained in Giesler (1982). Despite its partisanship, it contains the important documents pertaining to the case; and because of its partisanship it ought to be required reading for anyone desiring to defend evolutionary theory in the public forum.

5. There are of course crucial issues here for philosophers to vex. The view of "truth seeking" and "truth finding" in this book is that there are two fundamentally different contexts within which knowledge can be sought. (A) Knowledge can be sought within a determinate, historical investigative practice with fairly well-elaborated critical criteria of success and failure. Of course, such a practice is subject to historical transformations of several kinds. (B) Knowledge can be sought against the background of eternally threatening skepticism, and can only be achieved as eternal verities are discovered beyond the reach of the skeptic. Once achieved, these verities have no history. Alternative B will not be taken seriously. Within philosophy itself this is bound to look like an option for "internal realism," but if I were to address such issues here, I would claim that the presuppositions against which "internal realism" and "external realism" can be distinguished are themselves untenable. A number of current writers in the philosophy of science seem sympathetic to this line of thought. See, for example, Hacking (1981, 1983) and Fine (1984). People overwhelmingly tempted to talk of "relativism" note that the concept of scientific truth has become relative to place in an historical process. Such talk, however, betrays a nostalgia for the theological agenda taken over by secular philosophy: the search for some extra-historical van-

tage point (the god's-eye view), or for an autonomous philosophical enterprise having as its justifying task the search for such a vantage point. As Plato explicitly told us, the search for such a vantage point is a death wish. In the absence of the death wish, the term "relativism" is unavailable to do any useful work. It follows from my view that there is no point in vexing the issue of metaphysical realism. So I don't. This obviously provides a lovely purchase point for philosophers looking for one. Big deal.

6. Latour and Woolgar (1979/1986) should be studied quite carefully in connection with these last points.

Chapter 2

1. There is a recent object lesson in this regard, one that evolutionary theorists ought to think about carefully. It has not been smooth sailing through the critical seas for Brooks and Wiley's (1986) theory. I leave the bulk of the criticism of their theory in more capable hands. But as I pointed out in a review of the book (Dyke 1987), the ultimate source of many of the excesses of their theory lies in the challenge they pose for themselves. For they set the question, "Can evolutionary biology be a science?" They take this question to call for an evolutionary biology which is a deductive system in the old positivist sense. In particular, if evolutionary biology is to be a science, on their view, then systematics (undoubtedly in some cladistic form) must be a science. And this means that taxonomic trees (cladograms) must be derivable from appropriate first principles. This is a *very* strict requirement, one which many people (including me) think is farfetched given the complex multidimensional systems within which living things are imbedded. Yet, to say this *and* to accept their criteria of science entails that systematics will never be a science. There is, on the other side, a romantic wing of the systematists, stemming from the naturalist tradition. This wing apparently revels in the fact that systematics has not been made a science (especially in the deductive sense), and affirms its character as an art. In these terms, the view of this book would be that systematics cannot be a science in the deductive sense (very little is), but that the questions asked and answered by systematists can be asked and answered within science—understood as the science that has served us best and promises to continue to do so.

2. The demise of positivism as a viable philosophical view is by now a matter of history, and is often dated from the work of Kuhn and Lakatos. A full bibliography of the subsequent literature is clearly out of the question here. At the moment I find two works to be extremely useful benchmarks, one a "reply" to the other. The first is Bas Van Fraassen's *The Scientific Image* (Oxford 1980); the other is Churchland and Hooker's *Images of Science* (Chicago 1985). While polemically related, both books clearly show the degree to which the old positivist canon has been left behind. I will have further occasion to refer to particular contributions to the second work.

3. The notion of contrast space and the directly related notion of question space are not unique to Garfinkel. Peter J. Wilson (1983) uses the notion, as, in effect, does Pierre Bourdieu (1977). As Garfinkel points out, the notion of contrast space is implicit in the physicist's concept of state space; it is also implicit in many ways in economics through the notions of commodity spaces and production possibility surfaces.

I would also claim that the notion is implicit in Marx's writings, especially as a consequence of his concept of determination—but this is more controversial.

4. See Paul M. Churchland, "Cognitive Neurobiology: A Computational Hypothesis for Laminar Cortex," *Biology and Philosophy,* 1 (1) (1986).

5. A good example of this is the investigation of heterosis in corn. In searching for the source and meaning of heterosis, corn breeders are quite naturally led to concepts such as developmental homeostasis. But the corn breeder's work is deeply imbedded in a selection program as intensive and extensive as any ever undertaken. So in *this* case a selectionist baseline efficiently organizes research in a potentially profitable way. Notice, however, that it remains to be argued how the research done against this strongly selectionist background fits with the rest of the evolutionary picture. For example, the very concept of heterosis used by the corn breeders has but a tenuous

relationship to the concept of fitness as it is favored by more general evolutionary biology. The corn breeder identifies heterosis by means of a variety of yield measures, and these measures would have a wide variety of relations to the fitness of corn in any natural circumstances. Need we add that motilities within the corn genome complicate the picture still further (Mangelsdorf 1974; Iltis and Doebley in Grant 1983).

6. A succinct and illuminating parallel line of thought is that provided by Ronald N. Giere in "Constructive Realism" in Churchland and Hooker (1985).

7. The state space technique is suggested in Lewontin (1974), p. 8, as a useful tool in population biology. The difficulty with the suggestion, from my point of view, is that it is tied to a rigid conception of "predictive sufficiency," tying it to the old positivist program of explanation as derivation. The technique needn't be so tied. And see Rosen in Trappl (1986).

8. Compare Evelyn Fox Keller (1985) in her introduction to Part III.

9. Pauling and Zuckercandl in Fox and Dose (1977).

10. A fine recent discussion of the relevance of developmental constraints to evolutionary processes occurs in Maynard Smith et al. (1985).

11. Perhaps the clearest case of the establishment of an inertial baseline for social action is that of Ludwig Von Mises (1949). All social action is reduced to individual action; and individual action is reduced to want satisfaction. No wants, no action. I have discussed this view in a previous work (Dyke 1979).

12. See Garfinkel 1981, chapter 3.

13. For perhaps the most sophisticated dualism currently being discussed see Margolis (1983) and other of his contributions to the metaphysical debate.

14. We should not think that this can all be cashed in terms of the language of necessary conditions *unless* we notice that this language is appropriate only when a total set of theoretical and practical background conditions can be specified. The language of necessary and sufficient conditions is the mate to the rationalist dream of explanation as derivation, and the conditions for derivation are intolerably rigid, as we have seen, which is why I am uneasy with Lewontin's talk of "predictive sufficiency."

15. Parallel to Giere's characterization of his theory as "constructive realism" and the "constructivism" advocated by Sidney Fox and those associated with him, as well as Wicken, Bocchi and Ceruti, and Sermonti, I think my view can be called "constructive materialism" to distinguish it from the reductive varieties.

Chapter 3

1. I will follow lines of thought laid out by Roy Bhaskar (1975, 1979). From Bhaskar I take the rubric synchronic reduction/diachronic materialism and some thoughts concerning closure. A view similar to the one presented here appears in "Evolution of Levels of Evolution" by T. O. Fox, in Rohlfing and Oparin (1972). Prigogine and Stengers (1984) proceed along a parallel track that is very satisfying to me.

2. In each case I will attempt to preserve Garfinkel's insights about levels of explanation.

3. The usual thesis of reduction is that phenomenon A can be reduced to reduction base B if a complete (in the relevant sense, but see Garfinkel [1981]) account of A can be given entirely in terms of the theoretical apparatus of B. The theoretical apparatus of the reduction base is usually considered to consist of an inventory of entities and the laws governing their behavior. The reduction thesis is, then, that no reference to anything except the entities and laws of B is necessary to account for A. Reduction claims are defeated when it can be shown that accounts of A require (1) ineradicable reference to entities not in B, or (2) ineradicable invocation of laws not in B. Classic reduction claims (all of which seem to fail) are, in physics, the reduction of thermodynamics to statistical mechanics; in philosophy, the reduction of "mind" to "body"; and in social theory, the reduction of all social causation to economic causation.

In their canonical statement these are claims for *synchronic* reduction, in the sense that the description of the state of A at a particular time is meant to be a description of a state of B at the same time. The usefulness of the concept of *diachronic* reduction arises only under conditions of nonintegrability, and where synchronic reduction can be shown to fail, that is, when entities and/or law-bound connections have evolved in an ineradicably time-dependent way from prior states of systems accounted for (eventually) in the reduction base. If Prigogine and Stengers are right (op. cit.), then the evolution of the physical universe is a case involving diachronic reduction (as a matter of explanation). If I am right (Dyke 1983) then economic systems are another.

In practice, failures of synchronic reduction are signaled by discontinuities such as phase separations, bifurcations, radical shifts in reaction rate schedules (as in examples in the body of this chapter), or the breakdown of identity criteria (as in the case of intensionality). When a persistent discontinuity thwarts synchronic reduction it is precisely the task of a diachronic materialism to account for the discontinuity.

4. See Prigogine and Stengers (1984), chapter III, where it is noted that this observation is at least as old as Diderot. Also, ibid., p. 60, "The generality of dynamic laws is matched by the arbitrariness of the initial conditions." Ian Hacking has made similar observations, as has Nancy Cartwright (1983).

5. Pattee (1970) writes,

> In order to see the central problem of hierarchical organization more clearly, it is helpful to look at the difficulties which arise when the hierarchical interface is viewed from only one side or the other. Viewed from the lower side of this interface, the elementary laws are regarded as the given conditions and the problem is to see how the hierarchical constraints arise to perform integrated function at the higher level. Viewed from the upper side of the interface, the hierarchical constraints are regarded as the given conditions and the problem is to see if the integrated function is consistent with the elementary laws. (p. 121)

and

> To achieve function by permanently removing degrees of freedom in a collection of elements would be impossible. Instead the collection must impose *variable* constraints on the motion of individual elements. In physical language these amount to time-dependent boundary conditions on selected degrees of freedom. Furthermore, the time dependence is not imposed by an outside agent, but is inseparable from the dynamics of the system. Such constraints are generally called non-holonomic (non-integrable), and have an effect which is like modifying the laws of motion themselves. (p. 127)

6. For convenience I use Eigen et al. (1981). More substantial accounts appear in Eigen and Schuster (1983), and Kuppers (1983).

7. In fact it looks as though the hypercycle account will move back into a secondary position within a more robust synthesis. See the discussions in Matsuno et al. (1984) and Wicken (1985). The more "minimalist" the hypercycle account proves to be, the better it serves my purposes.

8. Here we have to be very careful not to confuse "stability" and "equilibrium." The cell is a "far from equilibrium" system. In addition, the assessment of thermodynamic efficiency is more complex than this sketch implies (Wicken 1987).

9. It is on issues such as these that the "worst case scenario" may lead us astray. Alternative theories are far less dependent on the "information hierarchy" of polynucleitide primacy. For example, a more robust claim would be that Haldane's soup was not homogeneous, but, rather, very "lumpy"—composed, in effect, by phase-separated subsystems. The further claim would be that the presence of phase separation is itself an essential part of the story of molecular evolution right from the beginning. As I suggested earlier, a story of molecular evolution along these lines would make my points even stronger than they are on the worst case scenario—in the sense that it assumes *minimum* structured structuring structures.

10. This point is directly related to the distinction between reversible and irreversible processes as they concern physicists, hence related to the issue of integrability and the search for timeless laws. Prigogine and Stengers (1984) discuss this extensively and intensively.

11. The distinction between bookkeeping and explanation—which will be exploited frequently over the course of this book—has somehow become common coin among philosophers of biology. I suspect that the remote source of the distinction is either Richard Lewontin (1974) or William Wimsatt. If Wimsatt did not invent the distinction, he is surely responsible for most of its profitable use.

Chapter 4

1. My conviction that a focus on closure conditions is essential has three sources: Garfinkel (1981), obviously; the work of Roy Bhaskar (1975, 1979); and a line of thought developed by Prigogine (1947), etc., Pattee (1970, 1972, 1973, 1978), P. A. Weiss (1973), and others.

2. See, for example, Trefil (1983), where, for example, gauge invariance is discussed. The moves required to argue for the appropriate closure in non-mathematical terms are very subtle (and, in this case, not very convincing). The parallel arguments within the mathematical formulation are another matter. The ultimate closure condition here is that you are on the right mathematical track.

3. For example, "Often the species that have strong effects on the population of other species will be the same as the species critical in the evolutionary unit of interaction, but this will not always be the case. For example, some mutualisms may have no effect on the population levels of interacting species. A mutualism is favored by selection because it allows those individuals possessing traits that foster the interaction to increase their genetic contribution to future generations relative to other individuals in the population. The mutualism may have little or no effect on the overall population levels of the species. Experiments designed to determine the unit of interaction within which selection acts significantly on all the species cannot use changes in population levels resulting from a manipulation of one of the species as the sole criterion of the limits of the important species." (Thompson 1982, p. 126) Notice that this important caution is necessary *even though* the criterion of inclusive fitness is accepted.

4. Issues such as these are discussed with great care in virtually all the papers collected in Futuyma and Slatkin (1983). *A propos* the present point, the editors say, "The study of coevolution forces a different view of genetic evolution than is usually adopted. In population genetics and evolutionary theory, each species is usually considered in isolation, with the environment and associated species relegated to the background, which is assumed to remain unchanged. Coevolutionary theory . . . assumes that genetic changes may occur in all interacting species, allowing genetic changes to be driven both by immediate interactions and by the feedback through the rest of the community. The distinctive feature of coevolution is that the selective factor (e.g., a predator) that stimulates evolution in one species (e.g., a prey) is itself responsive to that evolution, and the response should be predictable. In some cases a coevolutionary equilibrium may be established. In other cases there may be no coevolutionary equilibrium, and evolution may continue over longer time scales than are typical for the attainment of gene frequency equilibria as usually treated in population genetic models." (p. 6)

5. A model study of human populations is Friedlaender (1975). The conclusions of the study are exceedingly modest—simultaneously a tribute to scholarly care and the limitations of the research program.

6. A set of conspicuous exceptions can be found in Hochachka and Somero (1984). The authors are very scrupulous about providing closure conditions. Furthermore, their main topic, metabolism, lends itself to precise comparisons in terms of energetics.

7. Compare Futuyma and Slatkin (1983), p. 9, "Pimentel and his colleagues . . . have found evidence of genetic changes in both houseflies *(Musca)* and the parasitoid wasp *Nasonia vitripennis* when cultured together, and Hassell and Huffnaker (1969) reported increased resistance in the host and increased effectiveness of the parasite in a moth-wasp laboratory system. Such studies show, of course, that pair-wise coevolution is possible, not that it commonly occurs in nature. In the absence of an actual history of the dynamics of genetic change, the demonstration that each of two interacting species is genetically variable for the characteristics that affect their interaction can at least show the potential for coevolution." That is, they show the availability of the possibility space, but do not show that the space will necessarily be exploited.

Chapter 5

1. I used to talk of hierarchies all the time with a perfectly clear conscience. But Evelyn Keller changed all that.

2. Salthe (1985).

3. We ought to recall here the passage by Pattee quoted in chapter 4 (p. 146).

4. Pattee is again important here, as is the corresponding discussion in Garfinkel (1981).

5. But it seems to me that Wicken (1987) is, at the very least, a solid foundation for such a view.

6. The notion of genotype as phenotype is one with which John Jungck has been working profitably. See Jungck (1972).

7. In this regard, see Salthe's citation of K. N. Waltz, *The Theory of International Politics*, Addison Wesley, Reading, Mass., 1979.

8. The tension is clear in the following: "Hence the hierarchy of nature, while completely ordered with respect to composition (the constraint of nestedness), is only partially ordered with respect to the effects of processes. Transitivities across levels (Wimsatt's interactional complexity) do occur, both continually and intermittently." (Salthe 1985, p. 136).

9. Salthe is not crystal clear about this place or its consequences. He says, for example, "In order not to be misled by our own constructions, our representations of the world should always reflect our reference frame and not be constructed as if by an omnipresent observer. . . . " (p. 163), but then goes on to provide a heuristic for the construction of such a representation which proceeds from an initial "idealistic, observer-free model of the hierarchy of nature" through "a consideration of our finite, idiosyncratic perceptual abilities" to a local description of the hierarchy, and "Generalization from this particular local description of the hierarchy allows it to be transformed into *any* local description" (pp. 164–165). My worries about the first and last stages of the heuristic can be inferred from my remarks in earlier chapters.

10. But only in the sense at issue here. The same need not be true in terms of either thermodynamic efficiency in the classic Gibbsian sense, nor in terms of some Shannon-like measure of informational "entropy." The fact that a number of different conceptions of entropy occur in recent literature, but are not fully sorted out and related, is a serious impediment to achieving full clarity. The next few years should see some order being made out of the current chaos. Meanwhile, someone in my position of relative ignorance has to be extremely careful.

Chapter 6

1. A full set of references is not really necessary here. A short catalogue of the ways in which the nature/convention distinction has been made can be found in Park (1985).

2. This is found particularly in stoicism. E.g. " . . . the Stoics were the first to work out a detailed physical system based on the notion of a continuum in which all the parts intercommunicate. The conception of *pneuma,* the theory of total mixture or interpenetration, *krasis,* the doctrine of *sympatheia,* the conception of space and time as continua, all form part of a carefully elaborated and remarkably consistent whole. Their conception of the chain of cause and effect followed from this theory. Their belief in the possibility of predicting the future is, in turn, consistent with, and indeed entailed by, their determinism." (Lloyd 1973, p. 30)

Chapter 7

1. The classic discussions of these concepts are still those of J. L. Austin (1961, 1962).

2. I have argued that these problems extend to the behavior of people in the market (Dyke 1983). Garfinkel's concept of contrast spaces is particularly important when problems of intensionality arise. Naturalists among you might think how the following could be true: You don't sniff flowers because they're sex organs, but it's because they're sex organs that you sniff them.

3. The currently prevailing account of convention within analytic philosophy is that of David Lewis (1969). The criticisms of Lewis I would want to make—insofar as implied criticisms do not occur throughout my discussion—are conveniently made by Richard Shusterman (1986).

4. According to a well-established usage, "syntactics" refers to the rules of combination of linguistic symbols (words, etc.); "semantics" refers to the relation between the symbols and what they refer to (or how they are interpreted on a model); and "pragmatics" refers to the relation between the symbols and the symbol users.

5. This is another place where the discussion called for continues elsewhere—for example in the works of Pierre Bourdieu and others in his research group (cf. Bourdieu 1977, 1984). The point is that historical studies which would demonstrate what I only suggest here cannot very well be interpolated at this point. They are, however, beginning to become available.

6. In fact, it might not be a bad idea to include a key passage from the *Outline* here for comparative reference.

> The structures constitutive of a particular type of environment (e.g. the material conditions of existence characteristic of a class condition) produce *habitus,* systems of durable, transposable *dispositions,* structured structures predisposed to function as structuring structures, that is, as principles of the generation and structuring of practices and representations which can be objectively "regulated" and "regular" without in any way being the product of obedience to rules, objectively adapted to their goals without presupposing a conscious aiming at ends or an express mastery of the operations necessary to attain them and, being all this, collectively orchestrated without being the product of the orchestrating action of a conductor.
>
> The word *disposition* seems particularly suited to express what is covered by the concept of habitus (defined as a system of dispositions). It expresses first the *result of an organizing action,* with a meaning close to that of words such as structure; it also designates a *way of being,* a *habitual state* (especially of the body) and, in particular, a *predisposition, tendency, propensity,* or *inclination.* (p. 72)

7. A history of Jewish assimilation, and struggle against assimilation, written from this point of view would be especially welcome and instructive. Irving Howe (1976) has done work of the sort needed.

8. Hume says (1958, p. 32) "'Tis an establish'd maxim in metaphysics, *That whatever the mind clearly conceives includes the idea of possible existence,* or in other words, *that nothing we imagine is absolutely impossible."* Later (p. 267) he says "Nothing is more dangerous to reason than the flights of the imagination, and nothing has been the occasion of more mistakes among philosophers." Unfortunately he does not always take his own advice.

9. To embark on this line is *not* to reduce human practices and institutions to a utilitarian ethos. Shared tasks, mutually understood, need not have utilitarian ends. Here as with similar situations we have confronted, the claim that the utilitarian explanation is the preferred one depends on being able to enforce closure on explanatory space by means of the prior claim that the participants in the activity are at Malthusian-like limits. This may not be sustainable—even with respect to the lionesses.

Chapter 8

1. Latour and Woolgar (1979/1986) make this point particularly well. Hodgkins (1986) contains a convenient summary account of how the nematode C. elegans has become so well investigated that it can now serve as a tool for understanding developmental processes.

2. Richard Lewontin says in Maynard Smith et al. (1985), "evolution is best viewed as a history of organisms finding devious routes for getting around constraints."

3. Two representative sources for feminist thinking along these lines with particular reference to science are Harding (1986), and Bleier (ed.) (1986).

4. A brief but informative account of one women's studies program can be found in "Women's studies: They've come a long way to acceptance" by Lisa Foderaro (1985). This paper (and others) suggest strongly that the possibility space for women's studies programs varied greatly with respect to the affluence and prestige of the institution, and the corresponding affluence of its students. Students at Brown perceive themselves to have the leisure to spend part of their time in university on issues such as women's studies. Far fewer students at a place like Temple perceive themselves to have the time. The consequences for enrollment and stability are obvious.

5. Many current techniques, e.g. multivariate analysis, for dealing with complexity are inadequate—largely because of the additivity assumptions they make. There is a running critique of such techniques in Bourdieu (1984).

6. See, for example, Scott Gilbert, "Cellular Politics: Goldschmidt, Just, Waddington, and the Attempt to Reconcile Embryology and Genetics" (forthcoming); and Haraway (1976). To see how (substantially) the same point in research can be reached by different pathways, see, for example, Burian, Gayon, and Zallen (1986).

7. For this point, applied to the Skinnerian program, see Schwartz, Schuldenfrei, and Lacey, "Operant Psychology as Factory Psychology," *Behaviorism* (1978), 6, 229–254.

8. This point is made in one way or another by nearly every one of the contributors to Bocchi and Ceruti (1985). Ceruti's contribution, "La Hybris dell'Onniscienza e la Sfida della Complessita'" provides a convenient general statement.

9. Brandon (1985), following a suggestion of Wesley Salmon's, treats similar cases in terms of the concept of "screening off." We may also remember here Pattee's remarks, quoted earlier, about the ways in which higher level constraints modify underlying laws.

Chapter 9

1. In my judgment, the current most solid source is by Jeffrey Wicken (1985). The paper contains the most useful conceptual structure to date; is free of the outlandish claims that tend to infect the literature; and canvasses previous contributions in a productive way. The most important earlier contributions are, in addition to the key works developing NET in strictly biological contexts, Adams (1982); Boulding (1970); Odum (1971); and Proops (1983). Wicken (1985) provides a clear programatic statement to which I subscribe:

> All natural organizations . . . are products of evolutionary history. Since the functional referents of their operations are perpetuation and propagation, existence and operation are inseparable. This coupling requires that all natural organizations be *informed dissipative structures,* the integrity of whose organizational relationships depends on degrading free energy and dissipating entropy to a sink of some sort. The natural organization is inseparable from nature's overall dynamics.
>
> Socioeconomic systems belong to this class of organizations. While planning and design are essential to socioeconomic development, this development is of an evolutionary nature—bound by the hand of history on one hand, and the ecological-societal tolerances of what can work in sustaining organization on the other.

2. This allows us to link the present discussion to the discussion of emergence in chapter 3, hence directly to the context of the origin of life.

3. See Brooks and Wiley (1986).

4. Neither the utility nor the necessity of non-equilibrium thermodynamics is uncontested. See, for example, the Olympian review of Prigogine and Stengers (1984) by Heinz Pagels in *Physics Today,* January 1985. To a large extent the proof of the pudding will be in the development of the theory in the next few years, particularly in its ability to generate analytical data. More is already available than Pagels acknowledges. But, in addition, the acceptance of the view requires some very serious changes of metaphysical presupposition, of the sort we have been dealing with since the beginning of this book; and committed reductionists will undoubtedly hold out for as long as they can against it. I think it is fair enough for Pagels to be dubious at the present time. The proof of *his* pudding will be the cost he eventually has to pay for his reductionism.

5. These relationships may be far from simple. Multidimensionality makes the job of, for example, determining information measures of organization a very complex one. See Prigogine (1955); Proops (1983); Wicken (1985).

Chapter 10

1. The discernment of this attitude, however, may be more a misapprehension on the part of social scientists than a correct reading of the intentions of the sociobiologists. Close reading

of Wilson (1978) and Lumsden and Wilson (1981), for example, shows that *in their programmatic passages* they favor integration with the social sciences. On the other hand, as they offer substantive explanations, they tend to move closer to a reduction of the social sciences to biology.

2. In doing so they link their theorizing with a research tradition which is dominant within Western economic theories, and which has been elaborated with more mathematical rigor than any alternative research tradition within economics. In their more familiar uses, general equilibrium models are to be contrasted with the Keynesian and monetarist models which have been, perhaps, more familiar in the context of the politico-economic debate in this country over the last three decades. I think we can work with Kelley and Williamson's model in the present context without running afoul of the major contention between advocates of the various sorts of models.

3. Adams (1982) deals with the British economy in NET terms and so affords an interesting comparison here. However, Adams focuses on Britain in the latter part of the nineteenth century, a period in which the system had matured and was lapsing into a species of steady state, so there is not an exact parallel with Kelley and Williamson.

Bibliography

Adams, R. N. (1975). *Energy and Structure: A Theory of Social Power,* University of Texas Press, Austin and London.

———. (1982). *Paradoxical Harvest,* Cambridge University Press, Cambridge.

Allen, T. H. F., and Starr, Thomas B. (1982). *Hierarchy: Perspectives for Ecological Complexity,* University of Chicago Press, Chicago.

Austin, J. L. (1961). *Philosophical Papers,* Oxford University Press, Oxford.

———. (1962). *How to Do Things with Words,* Oxford University Press, Oxford.

Axelrod, Robert (1984). *The Evolution of Cooperation,* Basic Books, New York.

Beatty, John (1980). "What's Wrong with the Received View of Evolutionary Theory," in Asquith and Giere (eds.), *PSA 1980,* vol. 2, Philosophy of Science Association: East Lansing.

Bhaskar, Roy (1975). *A Realist Theory of Science,* Books, Leeds.

———. (1979). *The Possibility of Naturalism,* Humanities Press, Atlantic Highlands.

Birke, Linda, and Silvertown, Jonathan (1984). *More than Parts: Biology and Politics,* Pluto Press, London & Sydney.

Bleier, Ruth (ed.) (1986). *Feminist Approaches to Science,* Pergamon Press, New York.

Bocchi, Gianluca, and Ceruti, Mauro (1985). *La Sfida della Complessita',* Giangiacomo Feltrinelli Editore, Milano.

Boulding, K. E. (1970). *Economics as a Social Science,* McGraw-Hill, London.

———. (1981). *Evolutionary Economics,* Sage Publications, Beverly Hills.

Bourdieu, Pierre (1977). *Outline of a Theory of Practice,* Cambridge University Press, Cambridge.

———. (1984). *Distinction: A Social Critique of the Judgment of Taste,* Harvard University Press, Cambridge.

Brandon, Robert (1978). "Adaptation and Evolutionary Theory," *Stud. Hist. Phil. Sci,* 9:181–206.

———. (1985). "Adaptation Explanations," in Depew and Weber (1985).

Brannigan, Augustine (1981). *The Social Basis of Scientific Discoveries,* Cambridge University Press, Cambridge.

Braudel, Fernand (1981/82/84). *Civilization and Capitalism 15th–18th Century* (3 vols.), Harper and Row, New York.

Brooks, Daniel R., and Wiley, E. O. (1986). *Evolution as Entropy,* University of Chicago Press, Chicago.

Burian, Richard (1981/82). "Human Sociobiology and Genetic Determinism," *Phil. Forum,* 13:43–66.

―――. (1983). "Adaptation," in M. Grene (ed.), *Dimensions of Darwinism,* Cambridge University Press, Cambridge.

―――. (1985). "On Conceptual Change in Biology: The Case of the Gene," in Depew and Weber (1985).

Campbell, John H. (1985). "An Organizational Interpretation of Evolution," in Depew and Weber (1985).

Cartwright, Nancy (1983). *How the Laws of Physics Lie,* Clarendon Press, Oxford.

Cavalli-Sforza, L. L., and Feldman, M. W. (1981). *Cultural Transmission and Evolution: A Quantitative Approach,* Princeton University Press, Princeton.

Churchland, Paul, and Hooker, C. A. (1985). *Images of Science,* University of Chicago Press, Chicago.

―――. (1986). "Cognitive Neurobiology: A Computational Hypothesis for Laminar Cortex," *Biol. Phil.,* 1:25–52.

Cipolla, Carlo M. (1980). *Before the Industrial Revolution,* W. W. Norton and Co., New York.

Collier, John (1985). "Entropy in Evolution," *Biol. Phil.,* 1:5–24.

Conant, James B. (1947/1951). *On Understanding Science,* Yale University Press, New Haven.

Dawkins, Richard (1976). *The Selfish Gene,* Oxford University Press, Oxford.

Depew, David J., and Weber, Bruce H. (1985). *Evolution at a Crossroads: The New Biology and the New Philosophy of Science,* Bradford Books, MIT Press, Cambridge.

―――. (1986). "Non-equilibrium Thermodynamics and Evolution: A Philosophical Perspective," *Philosophica,* 37:27–57.

Dreyfus, Hubert L. (1983). "Why Current Studies of Human Capacities Can Never Be Scientific," Cognitive Science Program, Institute of Human Learning, University of California at Berkeley.

Dyke, C. (1979). *Philosophy of Economics,* Prentice-Hall, Englewood Cliffs, N.J.

―――. (1983). "The Question of Interpretation in Economics," *Ratio,* 25, 1:15–30.

―――. (1987). Review of Brooks and Wiley (1986), *Canad. Phil. Rev.,* 7:185-187.

Eigen, Manfred, et al. (1981). "The Origin of Genetic Information," *Sci. Am.,* 244 (15): 88.

――― and Schuster, P. (1983). *The Hypercycle,* Springer Verlag, Heidelberg.

Feigle, Herbert (1958/1967). *The Mental and the Physical,* University of Minnesota Press, Minneapolis.

Fetzer, James H. (ed.) (1985). *Sociobiology and Epistemology,* D. Reidel Publishing Co., Boston.

Feyerabend, P. K. (1975). *Against Method,* NLB, London.

Fine, Arthur (1984). "And not anti-realism either," *NOUS,* 18:51–65.

Foderaro, Lisa (1985). "Women's Studies: They've come a long way to acceptance," *Brown Alumni Monthly,* 86 (1).

Foucault, Michel (1965). *Madness and Civilization,* Random House, New York.

―――. (1970). *The Order of Things,* Vintage/Random House, New York.

―――. (1977). *Discipline and Punish,* Vintage/Random House, New York.

―――. (1980). *Power/Knowledge,* Pantheon, New York.

Fougereau, Michel, and Stora, Raymond (1986). *Aspects cellulaires et moléculaires de la biologie du developpement.* Elsevier Science Publishers, Amsterdam, New York.

Fox, Ronald Forrest (1971). *J. Theor. Biol.,* 31 (1):43–46.

Fox, Sidney, and Dose, Klaus (1977). *Molecular Evolution and the Origin of Life,* Marcel Dekker, New York.

Frey, Richard L., and Truscott, Alan F. (eds.) (1964). *The Official Encyclopedia of Bridge,* Crown Publishers, Inc., New York.

Friedlander, Jonathan Scott (1975). *Patterns of Human Variation,* Harvard University Press, Cambridge.

Froehlich, Jeffrey E. (1984). "Catastrophe Theory, Irreversible Thermodynamics, and Biology," *Centennial Rev.,* 228–250.

Futuyma, Douglas J., and Slatkin, Montgomery (1983). *Coevolution,* Sinauer Associates, Sunderland.

Garfinkel, Alan (1981). *Forms of Explanation,* Yale University Press, New Haven.

Gatlin, Lila (1972). *Information Theory and the Living System,* Columbia University Press, New York.

Geisler, Normal L. (1982). *The Creator in the Courtroom,* Mott Media Inc., Milford, Mich.

Georescu-Roegen, Nicholas (1971). *The Entropy Law and the Economic Process,* Harvard University Press, Cambridge.

Gilbert, Scott (forthcoming). "Cellular Politics: Goldschmidt, Just, Waddington, and the Attempt to Reconcile Embryology and Genetics."

Gillespie, N. O. (1979). *Darwin and the Problem of Creation,* University of Chicago Press, Chicago.

Gould, S. J., and Lewontin, R. C. (1979). "The Spandrels of San Marco and the Panglossian Paradigm," *Proc. Roy. Soc. London* B, 205.

Grant, William F. (ed.) (1983). *Plant Systematics,* Academic Press, Toronto.

Hacking, Ian (1981). "Do We See through a Microscope," *Pacific Phil. Quart.,* 62:305–322.

————. (1983). "Experimentation and Scientific Realism," *Phil. Topics,* 13:71–87.

Hall, Peter (1984). *The World Cities* (3rd ed.), St. Martins Press, New York.

Hamilton, W. D. (1964). "The Genetical Theory of Social Behavior," *J. Theor. Bio.,* 7.

Haraway, Donna J. (1976). *Crystals, Fabrics and Fields,* Yale University Press, New Haven.

————. (1981). "The High Cost of Information in Biological Theory," *Phil. Forum,* 13:2–3.

Harding, Sandra (1986). *The Science Question in Feminism,* Cornell University Press, Ithaca and London.

Hochachka, Peter. W., and Somero, George N. (1984). *Biochemical Adaptation,* Princeton University Press, Princeton.

Hodgkins, J. (1986). "Course 6. Notes on developmental biology and genetics of the nematode Caenorhabditis elegans," in Fourgereau and Stora (1986).

Howe, Irving (1976). *World of our Fathers,* Touchstone, Simon and Schuster, New York.

Hume, David (1958). *A Treatise of Human Nature,* ed. Selby-Bigge, Oxford University Press, Oxford.

Jacobs, Jane (1970). *The Economy of Cities,* Vintage Books, New York.

————. (1984). *Cities and the Wealth of Nations,* Random House, New York.

Jungck, John R. (1972). "The thermodynamics of self-assembly: an empirical example relating entropy and evolution," in Rohlfing and Oparin (1972).

Keller, Evelyn Fox (1983). *A Feeling for the Organism,* Freeman, New York.

————. (1985). *Reflections on Gender and Science,* Yale University Press, New Haven.

Kelley, Allen C., and Williamson, Jeffrey G. (1984). *What Drives Third World City Growth?,* Princeton University Press, Princeton.

Kubat, Libor, and Zeman, Jiri (1975). *Entropy and Information in Science and Philosophy,* Elsevier Scientific Publishing, Amsterdam, New York.

Kuhn, Thomas S. (1970). *The Structure of Scientific Revolutions,* University of Chicago Press, Chicago.

Kuppers, B. O. (1983). *Molecular Theory of Evolution,* Springer Verlag, Heidelberg.

Lakatos, Imre, and Musgrave, Alan (1970). *Criticism and the Growth of Knowledge,* Cambridge University Press, Cambridge.

Latour, Bruno, and Woolgar, Steve (1979/1986). *Laboratory Life: The Construction of Scientific Facts,* Princeton University Press, Princeton.

Levins, Richard (1966). "The Strategy of Model Building in Population Biology," *Am. Scientist,* 54:421–431.

————. (1968). *Evolution in Changing Environments,* Princeton University Press, Princeton.

Lewis, David (1969). *Convention: A Philosophical Study,* Harvard University Press, Cambridge.

Lewontin, R. C. (1974). *The Genetic Basis of Evolutionary Change,* Columbia University Press, New York.

————. (1976) "Evolution and the Theory of Games," in M. Grene and E. Mendelsohn (eds.), *Topics in the Philosophy of Biology,* Reidel, Dordrecht.

Lloyd, G. E. R. (1973). *Greek Science After Aristotle,* W. W. Norton and Co., New York.

Lumsden, C. J., and Wilson, E. O. (1981). *Genes, Mind, and Culture: The Coevolutionary Process,* Harvard University Press, Cambridge.

Lugt, Hans J. (1985). "Vortices and Vorticity in Fluid Dynamics," *Am. Scientist,* 73:162–167.

Mahar, Dennis (ed.) (1985). *Rapid Population Growth and Human Carrying Capacity,* World Bank Staff Working Paper No. 690, Population and Development Series No. 15.

Mangelsdorf, Paul C. (1974). *Corn: Its Origin, Evolution, and Improvement,* The Belknap Press of Harvard University, Cambridge.

Margolis, Joseph, (1983). *Philosophy and Psychology,* Prentice-Hall, Englewood Cliffs, N.J.

Matsuno, Koichiro, Dose, Klaus, Harada, Kaoro, and Rohlfing, Duane L. (eds.) (1984). *Molecular Evolution and Protobiology,* Plenum Press, New York, London.

Maynard Smith, J., Burian, R., Kauffman, S., Alberch, P., Campbell, J., Goodwin, B., Lande, R., Raup, D., Wolpert, L. (1985). "Developmental Constraints and Evolution," *Quart. Rev. Biol.,* 60:265–287.

Mayr, Ernst (1977). "Darwin and Natural Selection," *Am. Scientist,* 65:321–327.

Mills, S. K., and Beatty, J. H. (1979). "The Propensity Interpretation of Fitness," *Phil. Sci.,* 46:263–286.

Nitecki, Matthew H. (ed.) (1982). *Biochemical Aspects of Evolutionary Biology,* University of Chicago Press, Chicago.

Odum, Howard T. (1971). *Environment, Power, and Society,* Wiley, New York.

———— and Odum, Elizabeth S. (1976). *Energy Basis for Man and Nature,* McGraw-Hill, New York.

————. (1983). *Systems Ecology: An Introduction,* Wiley, New York.

Pagels, Heinz R. (1985). "Is the Irreversibility We See a Fundamental Property of Nature?," *Physics Today,* January: 97–98.

Park, Ynhui (1985). "Nature and Culture," *Contemporary Phil.,* 10:15–17.

Pattee, H. H. (1970). "The Problem of Biological Hierarchy," in C. H. Waddington (ed.), *Towards a Theoretical Biology,* Aldine Publishing Company, Chicago.

————. (1972). "The Evolution of Self-simplifying Systems," in E. Laszlo (ed.), *The Relevance of General Systems Theory,* Braziller, New York.

————. (1973). *Hierarchy Theory,* Braziller, New York.

————. (1979). "The Complementarity Principle in Biological and Social Structures," *Am. J. Physiol.* 236:R241–R246.

Peacocke, A. R. (1983). *An Introduction to the Physical Chemistry of Biological Organization,* Clarendon Press, Oxford.

Prigogine, Ilya (1955). *Introduction to the Thermodynamics of Irreversible Processes,* Wiley, New York.

———— and Stengers, Isabelle (1984). *Order out of Chaos.* Bantam Books, Toronto, New York.

Prodi, Giorgio (1977). *Le Basi Materiali della Significazione,* Bompiani, Milano.

Proops, J. L. R. (1983). *J. Soc. Biol. Struct.,* 6:353–362.

Rohlfing, Duane, and Oparin, A. I. (eds.) (1972). *Molecular Evolution,* Plenum Press, New York.

Rosen, Robert (1986). "The Physics of Complexity," in Trappl (1986).

Rosenberg, Alexander (1985). *The Structure of Biological Science,* Cambridge University Press, Cambridge.

Ruse, Michael (1971). "Natural Selection in the *Origin of Species,*" *Stud. Hist. Phil. Sci.,* 9:311–351.

————. (1975). "Charles Darwin's Theory of Evolution: An Analysis," *J. Hist. Biol.,* 8:219–241.

Saalman, Howard (1968). *Medieval Cities,* Braziller, New York.

Salthe, Stanley N. (1985). *Evolving Hierarchical Systems: Their Representation and Structure,* Columbia University Press, New York.

Scargill, D. I. (1979). *The Form of Cities,* St. Martins Press, New York.

Schroedinger, E. (1945). *What is Life?,* Cambridge University Press, Cambridge.

Schwartz, Barry, Schuldenfrei, Richard, and Lacey, Hugh (1978). "Operant Psychology as Factory Psychology," *Behaviorism,* 6:229–254.

————. (1986). *The Battle for Human Nature,* W. W. Norton and Co., New York and London.

Shusterman, Richard (1986). "Convention: Variations on a Theme," *Phil. Investigations,* 9:36–55.

Simon, Herbert A. (1981). *The Sciences of the Artificial,* MIT Press, Cambridge.

Slobodkin, Lawrence B. (1961/1980). *Growth and Regulation of Animal Populations,* Dover, New York.

————. (1968). "Towards a Predictive Theory of Evolution," in R. Lewontin (ed.), *Population Biology and Evolution,* Syracuse University Press, Syracuse.

Smith, Joseph Wayne (1984). *Reduction and Cultural Being: Antireductionist critique of positivist programs,* Martinus Nijhoff, The Hague.

Sober, Elliot, and Lewontin, R. C. (1982). "Artifact, Cause, and Genic Selection," *Phil. Sci.,* 49:157–180.

Stoianovich, Traian (1976). *French Historical Method: The Annales Paradigm,* Cornell University Press, Ithaca and London.

Thompson, John M. (1982). *Interaction and Coevolution,* Wiley, New York.

Trappl, Robert (ed.) (1986). *Power, Autonomy, Utopia: New Approaches toward Complex Systems,* Plenum Press, New York.

Trefil, James S. (1983). *The Moment of Creation,* Collier Books, MacMillan, New York.

Tuomi, J., and Hauoija, E. (1979). "An Analysis of Natural Selection in Models of Life-history Theory," *Savonia,* 3:9–16.

————. Sala, Jukka, Haukioja, Erkki, Niemela, Pekka, Hakala, Tuomo, and Mannila, Rauno (1983). "The Existential Game of Individual Self-maintaining Units: Selection and Defense Tactics of Trees," *Oikos,* 40:369–376.

Van Fraasen, Bas (1980). *The Scientific Image,* Oxford University Press, Oxford.

Von Mises, Ludwig (1949), *Human Action,* Yale University Press, New Haven.

Weaver, Warren (1948). "Science and Complexity," *Am. Scientist,* 36:536–540.

Weber, Max (1947). *The Theory of Social and Economic Organizations,* Oxford University Press, New York.

Weiss, P. A. (1973). *The Science of Life,* Futura Publishing, Mt. Kisco, N.Y.

Wicken, J. S. (1979). "The Generation of Complexity in Evolution: A Thermodynamic and Information-theoretical Discussion," *J. Theor. Biol.,* 77:349–365.

————. (1980). "A Thermodynamic Theory of Evolution," *J. Theor. Biol.,* 87:9–23.

————. (1985). "An Organismic Critique of Molecular Darwinism," *J. Theor. Biol.,* 117:545–561.

————. (1986). "Evolutionary self-organization and entropic dissipation in biological and socioeconomic systems," *J. Soc. Biol. Struct.,* 9:261–273.

————. (1987). *Evolution, Thermodynamics, and Information,* Oxford University Press, New York.

Wilson, E. O. (1971). *The Insect Societies,* Harvard University Press, Cambridge.

————. (1975). *Sociobiology,* Harvard University Press, Cambridge.

————. (1978). *On Human Nature,* Harvard University Press, Cambridge.

Wilson, Peter J. (1983). *Man the Promising Primate,* Yale University Press, New Haven.

Wittgenstein, Ludwig (1953). *Philosophical Investigations,* MacMillan, New York.

Wolff, Robert Paul (1984). *Understanding Marx,* Princeton University Press, Princeton.

Index